Tesla Technology

Lost Inventions, Radio Tesla, Tesla Coil **merged and amplified**

by George Trinkaus

High Voltage Press

Portland

Cover illustration: Tesla's rotating brush, a high-voltage
brush discharge in a vacuum. See page 23.

design: Eberhardt Press

to Cora and Jessie

to Emily

Contents

Preface

The Tesla Technology Series, published in booklets for twenty-eight years (1986 to 2015), is finally combined into a single volume. This project is long overdue, because the booklets have continued to sell and to sustain their author, and also because no digital versions of these works existed that I could properly edit until 2013, when I scanned and OCR'd them into digital type. This labor completed, a new combined edition could be constructed, not as a potboiler patchwork of PDFs, but as a completely new work, expanded, reedited, reset, and redesigned.

I have edited the old works as fresh manuscript, making many new inserts, as well as correcting various errors which have been nagging me for decades. New to print, but previously posted on my Teslapress, are "Tesla Electrotherapy" and "The Tesla Mystique." I have added many new illustrations, which now total about 240.

About my qualifications: Except that I have built a Tesla coil, I have no special direct knowledge of Tesla. I never knew the man. I am not his "channel." This work is simply one person's distillation of the existing Tesla literature, his published notes and lectures, his magazine articles, as well as biographies and other secondary sources. But most of my writing energy has gone into translating into informal English the techno-legalese of Tesla's patents.

Tesla was eloquent in English (and several other languages as well). This shows in his patents, and I quote him extensively. But Tesla's patents, like all patents, make tough reading, because they are not written for the curious but are defensive, legalistic exercises designed to protect the inventor's interests. Many of the illustrations in this book are from Tesla patent drawings. In the original drawings, components had been identified with alphabetic letters or with numerals which are referred to in the patent text. Since these labels would mean nothing to you, I have taken the liberty of deleting them and substituting appropriate labels to identify the key components.

In the illustration you get the patent number so you can obtain the patent for further research. You also get the year of the patent. Note that this is not the year the patent was granted but the year Tesla applied for it. Sometimes there is a meaningful interval. In the case of the magnifying transmitter there is an interval of 13 years.

Like many of my readers, I have been fascinated with electricity since my youth. I was a pre-teen basement experimenter and a novice-class ham (WN3UFH). I read all available books on the subject. This exploration was interrupted in 1952 (in the vacuum-tube era), when, at age 15, I was sent off to prep school. My inquiry into the magic of electricity lay dormant for 35 years, until about 1984, when I encountered the Tesla patents.

Like Rip Van Winkle, I awoke to an electrical world completely changed. The gorgeous vacuum-tube had been replaced by the gray subtleties of solid state, Everything was miniaturized as if intended to fly into space on a rocket. The traditional unit of frequency, a perfectly sensible cycles-per-second (cps), was mysteriously replaced by the Hertz (Hz), a subtle insult to Tesla.

It seemed to be required that I re-educate myself into the electronic new world order, which I proceeded to do with some help from a friend.

In those dormant years, my electric interests persisted at a low ebb. At my liberal arts college, I took the one course offered in electronics. Out in the corporate world, where I somehow became an editor of textbooks, I was allowed to preside over the publication of a series of basic electronics books for the schools, perhaps because I was the only techno-literate person in the office. But, now, I confess that I never really understood how electricity works until I read Tesla. I had to de-school myself to write this book.

—GT

Tesla
a capsule biography

Tesla, Nikola (1856 - 1943), electrical inventor. Born Serbia. Educated at the polytechnical school at Graz and at University of Prague. Worked as telephone engineer in Prague and Paris. Conceived new type of electric motor having no commutator, as DC motors have, but works on principle of a rotating magnetic field produced by polyphase alternating currents. Constructed prototype. Found nobody interested in Europe.

Emigrated to U.S. (1884). Worked briefly and unhappily with Thomas Edison. Established own lab and obtained patents on polyphase motors, dynamos, transformers for a complete alternating-current power system. Formed alliance with George Westinghouse, who bought polyphase patents for $1 million plus royalty. With Westinghouse, engaged in struggle against Edison to convince public of the efficiency and safety of AC over DC. Succeeded in getting AC accepted as the electric power system worldwide. Also with Westinghouse, lit the Chicago World's Fair, built Niagara Falls hydro-power plant, and installed AC systems at Colorado silver mines, other industries. By turn of the century was lifted to celebrity status comparable to Edison's, as media promoted Tesla as the poster boy for an expanding electric power industry.

Experimenting independently in Manhattan lab, developed and patented electric devices based on the superior capabilities of high-potential, high-frequency currents: Tesla coil, radio, high-frequency lighting, x-rays, electrotherapy. Suffered lab fire. Rebuilt, continued.

Moved lab to Colorado Springs for about one year (1899). Built huge magnifying transmitter. Experimented with wireless power, radio, earth resonance. Studied lightning. Created lightning. Returned to New York. With encouragement of financier LP. Morgan, promoted a World System of radio broadcasting utilizing magnifying transmitters. Built huge tower for magnifying transmitter at Wardencliff, Long Island as first station in World System. Received enough from Morgan to bring station within sight of completion, then funds cut off, project collapsed.

Continued to invent into the 1920s, but flow of patents meager compared to earlier torrent, which amounted to some 700 patents worldwide. High-frequency inventions ignored by established technology, as were disk turbine, free-energy receiver, other inventions. Shut out by media except for birthday press conferences. At these predicted microwaves, TV, beam technologies, cosmic-ray motor, interplanetary communications.

In the 1930s involved in wireless power projects in Quebec. Last birthday media appearance in 1940. Died privately and peacefully at 87 in his rooms at the hotel New Yorker from no apparent cause in particular. Personal papers, including copious lab notes, impounded by U.S. Government, surfaced many years later at a Tesla Museum in Belgrade. Of these notes, only a fragment, Colorado Springs Notes, has been published by the Museum.

Part 1
Lost Inventions

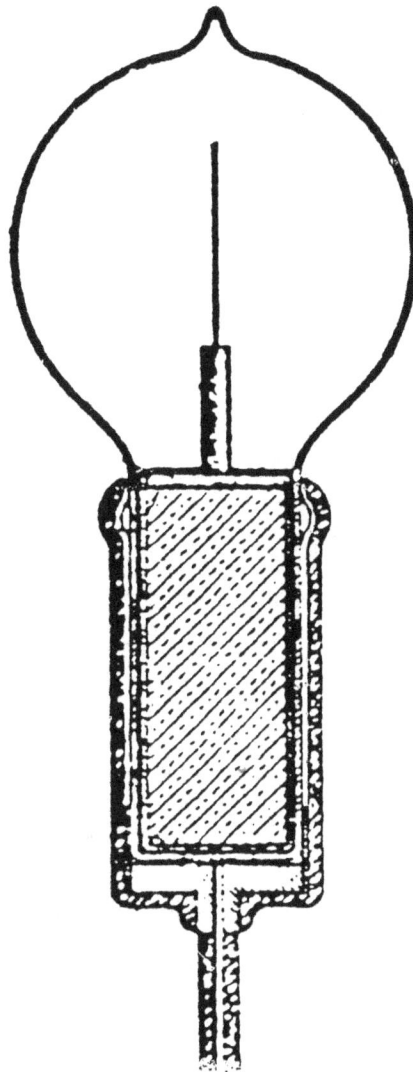

CHAPTER 1

Disk-Turbine Rotary Engine

Tesla called it a "powerhouse in a hat." One version developed 110 horsepower at 5000 RPM and was less than ten inches in diameter. Tesla believed larger turbines could achieve 1000 horsepower. The disk-turbine rotary engine runs vibration-free. It is cheap to manufacture because nothing but the rotor bearings must be fitted to close tolerances. It requires little maintenance. If necessary, the rotor can be replaced with ease. The turbine can run on steam, compressed air, gasoline, or oil.

how it works

Unlike conventional turbines that use blades or buckets to catch the flow, Tesla's uses a set of rigid metal disks that, instead of battling the propelling stream at steep angles, runs with smooth efficiency in parallel with the flow. What drives the disks is a peculiar adhesion that exists between the surface of a body and any moving fluid. This adhesion, a hindrance to aircraft and other vehicles, is caused by "the shock of the fluid against the asperities of the solid substance" (simple resistance) and "from internal forces opposing molecular separation" (a sticking phenomenon).

The propellant enters the intake and is nozzled onto the disks at their perimeter. It travels over the spinning disks in a spiral fashion, exiting at the disks' central openings where it is exhausted from the casing. Tesla notes in his patent that, in an engine driven by a fluid, "changes in the velocity and direction of movement of the fluid should be as gradual as possible." This, he observes, is not the case in existing engines where "sudden changes, shocks, and vibrations are unavoidable." Notes Tesla, "The use of pistons, paddles, vanes and blades necessarily introduces numerous effects and limitations and adds to the complication, cost of production, and maintenance of the machines." We who are stuck with the piston engine know this all too well. The Tesla turbine is vibration-free because the propelling fluid moves "in natural paths or stream lines of least resistance, free from constraint and disturbance." The turbine is easily reversed by conducting the propellant through the intake valve on the other side.

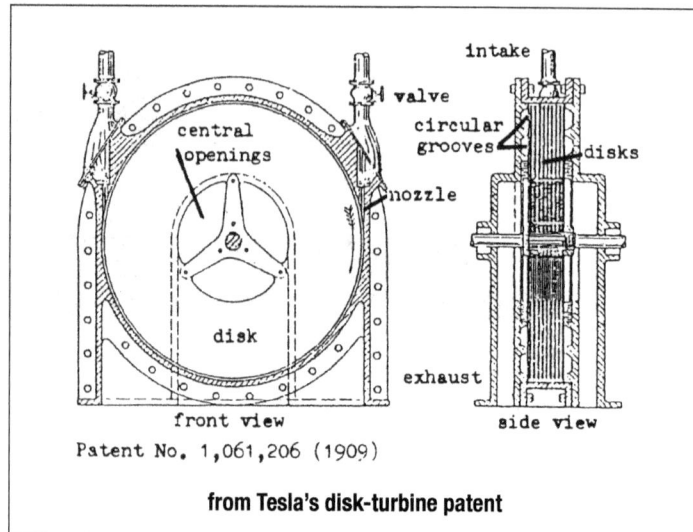

Patent No. 1,061,206 (1909)

from Tesla's disk-turbine patent

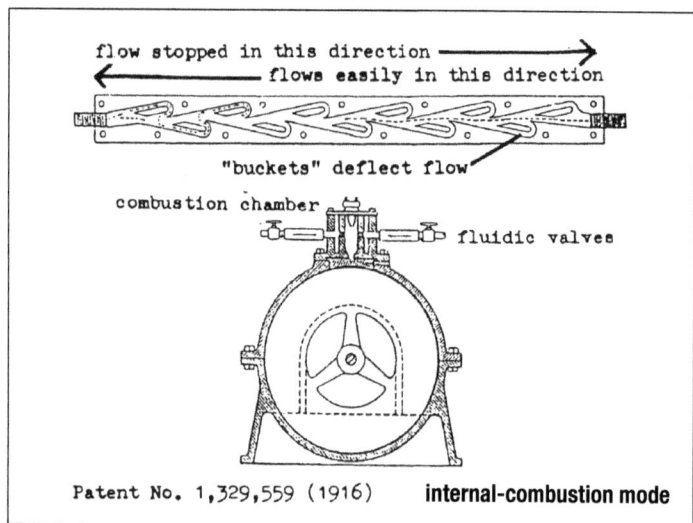

Patent No. 1,329,559 (1916)　**internal-combustion mode**

internal combustion

A hollow casting is bolted to the top of the turbine for the internal combustion mode. A glow-plug or sparkplug screws into the top of this chamber. Sticking out of the sides are the intake valves. Interesting thing about these valves: there are no moving parts. They work on a fluidic principle. The Tesla turbine's only moving part is its rotor. Imagine, a powerful internal combustion engine with only one moving part.

fluidics

The fluidic valve, which Tesla calls a valvular conduit, allows easy flow in one direction, but in the other direction the flow gets hung up in dead-end chambers (buckets) where it gets spun around 360 degrees, thus forming eddies, or counter-currents that stop the flow as surely as if a mechanical valve were moved into the shut position.

The spinning rotor creates plenty of suction to pull fuel and air into the combustion chamber. Tesla notes that "after a short lapse of time the chamber becomes heated to such a degree that the ignition device may be shut off without disturbing the established regime." In other words, it diesels. By running it in reverse, the disk-turbine motor becomes a very efficient pump. (Tesla's Patent No. 1,061,142).

fluid drive

The disk turbine principle Tesla also employed in the speedometer, which presents the problem of having to turn the rotary motion of a vehicle's wheels into angular motion in order to push a spring-loaded indicator needle over a short arc. Tesla's solution: the speedometer cable connects to a disk which spins in interface with a second disk, thus imparting spin to the fluid in between and, hence, to the second disk which moves the needle.

Interface two disks of different sizes in a fluid medium and "any desired ratio between speeds of rotation may be obtained by proper selection of the diameters of the disks," observes Tesla in his patent, thus anticipating in 1911 the fluid-drive automatic transmission.

Tesla first worked on his turbine early in his career, believing it would be a good prime mover for his alternating-current dynamos. He thought they would be far superior to the reciprocal steam engines that were the workhorses of that era. But he did not get down to perfecting and patenting his turbine until after the collapse of his global broadcasting scheme (1909). By this time the internal-combustion piston engine was firmly rooted in Western power mechanics.

Tesla referred to "organized opposition" to his attempts to introduce the superior engine, and so have others who have made the attempt since. But Tesla still saw a glorious future for his turbine. To his friend, Yale engineering professor Charles Scott, Tesla predicted, "My turbine will scrap all the heat engines in the world." Replied Scott, "That would make quite a pile of scrap."

CHAPTER 2

Spark-Gap Oscillator

Patent No. 462,418 (1891)

Tesla was central in establishing the 60-cycle AC power system still in use today. Yet he suspected that more striking phenomena resided in the higher frequencies of electric vibration. To reach these heights, he first tried dynamos spun at higher speeds and having a greater number of poles than any that had existed before. One version, which had as an armature a flat, radially grooved copper disk, achieved 30,000 cycles, but Tesla wanted to go into the millions of cycles.

It occurred to him that this vibratory capability was to be found in the capacitor. With a capacitor circuit, the spark-gap oscillator, he did indeed achieve the higher frequencies, and he did so by non-mechanical means. The circuit was promising enough for him to patent it as "A Method of and Apparatus for Electrical Conversion and Distribution," for Tesla saw in it the possibility of a whole new system of electric lighting by means of high frequencies.

Though it was quickly succeeded by the Tesla coil and is not numbered among the more famous of the lost inventions, the spark-gap oscillator is pivotal for Tesla as the invention that launched him into his career in high frequencies.

how it works: the capacitor

There are only a few basic building blocks of electrical circuitry. The capacitor is one of them. Tesla didn't invent it, it had been around for some time, arguably for millennia, but he did improve upon it in three of his patents.

Also called a condenser, the common capacitor is just a sandwich of conductive and nonconductive layers that serves the purpose of storing electrical charge. The simplest capacitor has

just two conductive sheets separated by a single sheet of insulation. In the capacitor shown, the conductive elements are two metal plates. The insulation between them is oil. In the official vocabulary, the plates are indeed called "plates" and the insulative layer (oil, glass, mica, or whatever) is called the "dielectric."

Connect the two terminals of a capacitor into a circuit where there is plus-minus electrical potential, and charge builds on the plates, positive on one, negative on the other. Let this charge build for an interval, then connect the two plates through some resistance, a coil, say, and the capacitor discharges. Very suddenly. Tesla said that "the explosion of dynamite is only the breath of a consumptive compared with its discharge." He went on to say that the capacitor is "the means of producing the strongest current, the highest electrical pressure, the greatest commotion in the medium."

The capacitor's discharge is not necessarily a single event. If it discharges into a suitable resistance, there is a rush of current outward, then back again, as if it were bouncing off the resistance, then out, and back and so forth until it diminishes to nothing.

The character of the vibration is determined in part by the capacity of the capacitor, that is, how much charge it will hold.

Patent No. 447,921 (1891)

high-frequency dynamo

Patent No. 464,667 (1891)

capacitor

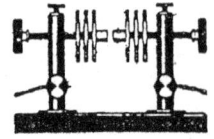

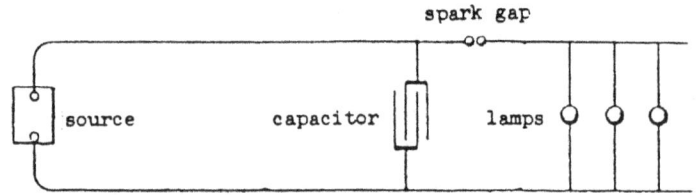

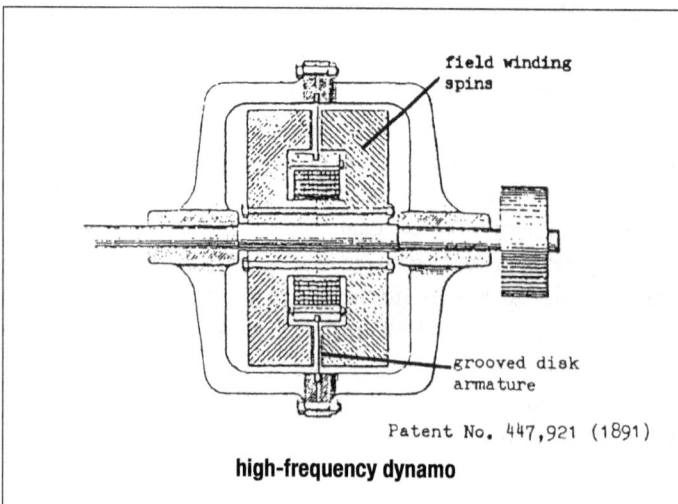

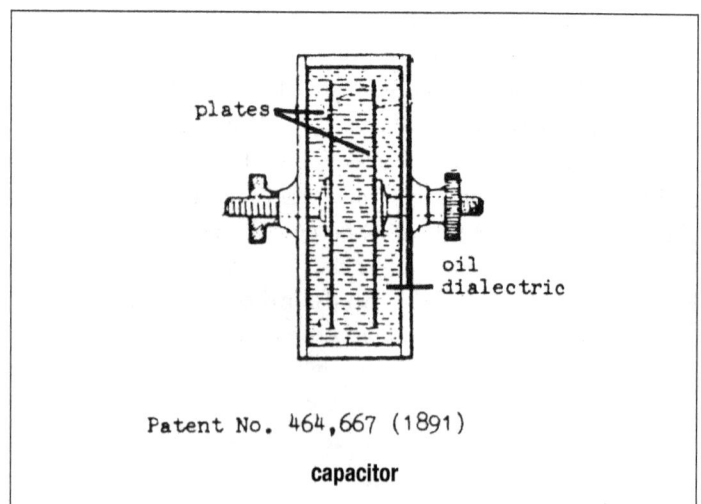

This is a function of its size, the distance between plates, and the composition of the dielectric. Upon discharge there would be, typically, a fundamental vibration, some harmonics, and perhaps other commotion, maybe musical, maybe not. Additional circuitry can tame the vibration to a "pure" tone.

the "medium"

When Tesla speaks of "commotion in the medium," what is the "medium?" In Tesla's time it was an article of faith that there existed a unified field that permeated all being called the "ether." The ether as the electric medium still is an article of faith in some circles, but in official science its existence is presumed to have been disproved in the laboratory. Nevertheless, this conviction about an ether ran very deep, not only among scientists but among all thinkers, until quantum theory, Einstein, and, finally Hiroshima firmly established the new faith.

Tesla said the electron did not exist, The materialistic concept of these little particles running over conductors is alien to Tesla electric theory.

Until relatively recently, an ether was taken as an article of scientific faith and of theological faith as well. Here is the Quaker writer Rufus Jones on the ether in 1920: "An intangible substance which we call ether – luminiferous (light-bearing) ether fills all space, even the space occupied by visible objects, and this ether which is capable of amazing vibrations, billions of times a second, is set vibrating at different velocities by different objects. These vibrations bombard the minute rods of the retina … It is responsible also for all the immensely varied phenomena of electricity, probably, too of cohesion and gravitation … The dynamo and the other electrical mechanisms which we have invented do not make or create electricity. They merely let it come through, showing itself now as light, now as heat, now again as motive power. But always it was there before, unnoted, merely potential, and yet a vast surrounding ocean of energy there behind, ready to break into active operation when the medium was at hand for it."

Jones, who was not a scientist but a religious thinker and communicator, was making a point about the nearness of God's power and could do so by invoking the physics of his time. This would be difficult using the quantum physics in fashion today, which W. Gordon Allen has called "atheistic science."

elastic

Although the ether is intangible, it is assumed to have elastic properties, so that Tesla can say "a circuit with a large capacity behaves as a slack spring, whereas one with a small capacity acts as a stiff spring vibrating more vigorously." This elastic character of the ether, which you experience palpably when you play with a pair of magnets, is due to the medium's lust for equilibrium. Distorted by electrical charge (or by magnetism or by the gravity of a material body), the ether seeks to restore a perfect balance between the polarities of positive-negative, plus-minus,

yang-yin. Voltage is the measure of ether strain or imbalance, called potential difference, or just potential.

Balance is not restored from this strained condition in one swing-back. As we have seen with the capacitor, the disturbed electric medium, like a plucked guitar string, over-swings the center line of equilibrium to one side, then to the other, again and again, and this we know as vibration.

In this way of looking at nature, vibration is energy, energy is vibration. So you could say that the commotion in the medium caused by the capacitor's discharge is energy itself. Thus, you can speak of the capacitor as an energy magnifier. Even though a feeble potential may charge it, the sudden blast of the capacitor's release plucks the medium mightily.

The capacitor is common in modern circuitry, but Tesla used it with much greater emphasis on its capability as an energy magnifier and on a scale almost unheard of today. It's difficult to find commercial capacitors that meet Tesla specifications. Builders of Tesla coils and other high-voltage devices usually must scrounge exotic surplus sources. So-called door-knob caps can be found rated 35 KV. Builders often construct their own capacitors. Fortunately, this can be done using readily available materials, as shown later in Part 3, Tesla DIY.

how it works: the spark gap

A simple way to discharge a capacitor is through a spark gap. The spark-gap oscillator is just a capacitor firing into a circuit load (lamps or whatever) through a spark gap. The opening between the spark-gap electrodes determines the threshold at which the capacitor will fire and hence the pulse rate. The frequency is determined by the reactance, or the bounce characteristics of the circuit.

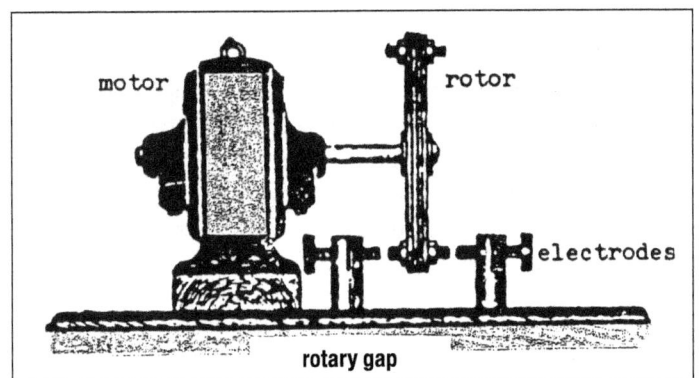

rotary gap

The potential needed to bridge the gap is in the tens of thousands of volts. It takes a potential of about 20,000 volts to break down the resistance of just a quarter inch of air. The gap doesn't necessarily have to be air. Tesla has referred to a gap consisting of a "film of insulation."

A spark gap is a switching device, a semiconductor, in fact. But the spark gap is problematic, particularly the common two-electrode air-gap version. Heating and ionizing of the air cause irregularities in conduction and premature firing.

Conduction lingers beyond the desired cut-off. This arcing must be quenched. It can be to a great degree by using a series of small gaps (series gap), or by using a rotary gap. Tesla also immersed the gap in flowing oil, used an air blow-out, and even found that a magnetic field helps to quench. For the gap Tesla substituted high-speed rotary switches which he called "circuit controllers." One has a rotor that dips into a pool of mercury, and another uses mercury jets to make contact.

You can operate a spark gap without a capacitor by connecting it directly to a source of sufficient voltage. This is of course how our automotive spark plugs work, directly off the ignition coil. (The capacitor in that circuit is used to juice the ignition coil primary.) The automotive distributor, incidentally, is a rotary circuit controller, pure Tesla.

Early radio amateurs used spark-gap oscillators as transmitters. The capacitor was, more often than not, left out of the circuit, but with it the transmitter could create a greater "commotion in the medium."

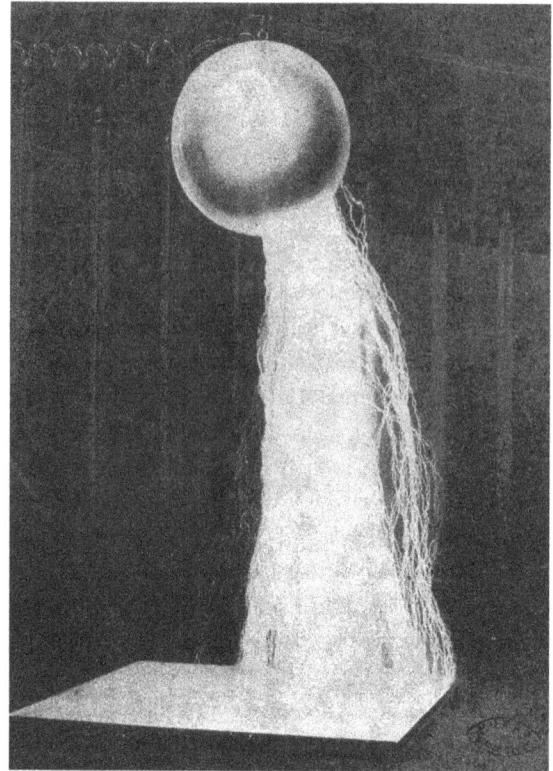

CHAPTER 3

Tesla Coil

Tesla's best-known invention takes the spark-gap oscillator and uses it to vibrate vigorously a coil consisting of few turns of heavy conductor. Inside of this primary coil sits another secondary coil with hundreds of turns of slender wire. In the Tesla-coil transformer there is no iron core as in the conventional step-up transformer, and this air-core transformer differs radically in other ways.

Recounting the birth of this invention, Tesla wrote, "Each time the condenser was discharged the current would quiver in the primary wire and induce corresponding oscillations in the secondary. Thus evolved a transformer or induction coil on new principles. Electrical effects of any desired character and of intensities undreamed of before are now easily producible by perfected apparatus of this kind." Elsewhere Tesla wrote, "There is practically no limit to the power of an oscillator."

The conventional step-up transformer (short primary winding, long secondary on an iron core) boosts voltage at the expense of amperage. This is not true of Tesla's transformer. There is a real gain in power. At his Colorado Springs lab, Tesla experimented with coils having outputs in excess of 12 million volts. He wrote, "It was a revelation to myself to find out that a single powerful

streamer breaking out from a well insulated terminal may easily convey a current of several hundred amperes! The general impression is that the current in such a streamer is small."

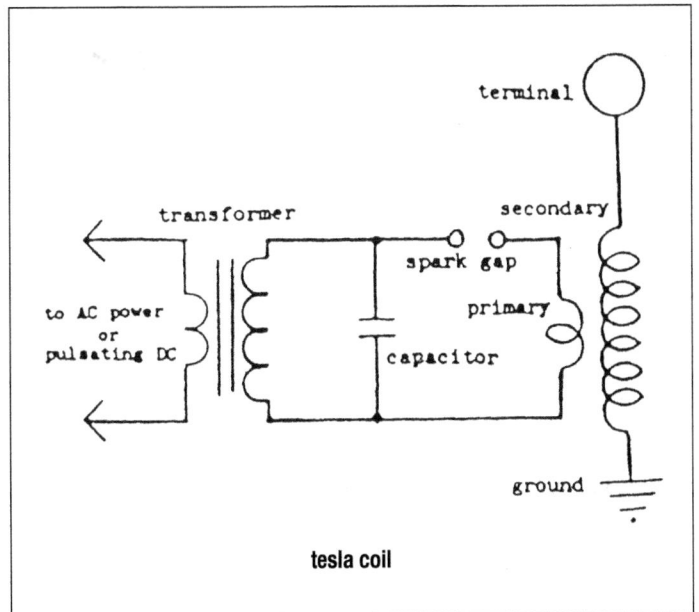

tesla coil

how it works

A Tesla-coil secondary has its own particular electrical character determined in part by the length of that slender coiled wire. Like a guitar string of a particular length, it wants to vibrate at a particular frequency. The secondary is inductively plucked by the primary coil. The primary circuit consists of a pulsating high-voltage source (a generator or conventional step-up transformer), a capacitor, a spark gap, and the primary coil itself. This circuit must be designed so that it vibrates at a frequency compatible with the frequency at which the secondary wants to vibrate.

The primary circuit's frequency is determined by the reactance of the circuit, including the capacity of the capacitor and the reactive character of the primary coil, determined in part by the length of its winding. The setting of the spark-gap is a factor in the performance of the primary circuit. When all of these primary-circuit components are tuned to work in harmony with each other, and the circuit's resulting frequency is right for plucking the secondary in a compatible rhythmic manner, the secondary becomes at its terminal end maximally excited and develops huge electrical potentials, which if not put to work, boil

off as a corona of bluish light or as sparks and streamers that jump to nearby conductors with crackling reports. Build a Tesla coil. See Part 3, Tesla DIY.

Unlike the conventional iron-core step-up transformer, whose core has the effect of damping vibrations, the secondary of the Tesla transformer is relatively free to swing unchecked. The pulsations from the primary coil have the effect of pushing a child in a swing. If it's done in a rhythmic manner at just the right moment at the end of a cycle, the swing will oscillate up to great heights. Similarly, with the right timing, the electrical vibration of the secondary can be made to swing up to tremendous amplitudes, voltages in the millions. This is the power of resonance.

man-made earthquake

Tesla was fascinated with the power of resonance and experimented with it not only electrically but on the mechanical plane as well. In his Manhattan lab in what is now called Soho, he built mechanical vibrators and tested their powers. One experiment got out of hand. To a steel pillar Tesla attached a powerful little vibrator driven by compressed air. Leaving it there, he went about his business. Meanwhile, down the street, a violent quaking built up, shaking down plaster, bursting plumbing, cracking widows, and breaking heavy machinery off its anchorages.

Tesla's vibrator had found the resonant frequency of a deep sandy layer of subsoil beneath his building, setting up an earthquake. Soon Tesla's own building began to quake, and, as the story goes, just at the moment the police burst into the lab, Tesla was seen smashing the device with a sledge hammer, the only way he could promptly stop it.

In a similar experiment, on an evening walk through the city, Tesla attached a battery-powered vibrator, described as being the size of an alarm clock, to the steel framework of a building under construction and, adjusting the vibrator to a suitable frequency, set the structure into resonant vibration. The structure shook, and so did the earth under his feet. Later Tesla boasted that he could shake down the Empire State Building with such a device, and, as if this claim were not extravagant enough, he went on to state that a large-scale resonant vibration was capable of "splitting the Earth in half."

Tesla's vibrators are not in patent as such but probably resembled his reciprocating engines (such as Patent No. 511,916). These exploited the elasticity of gases, as his Tesla coil exploits the elasticity of the electric medium.

a new power system

Tesla invented his resonant transformer, as the Tesla coil is sometimes called, to power a new type of high-frequency lighting system, as his 1891 patent drawing shows. This was the first Tesla-coil patent. There followed a series of other patents developing the device. All of these are for bipolar coils: both ends of the secondary are connected to the working circuit (usually lamps), as opposed to the monopolar format favored by today's basement builders, in which the top of the secondary is connected to a ball or other terminal capacitor, the bottom to ground. The monopolar format emerges later in patents for radio and wireless power, including Tesla's magnifying transmitter.

The 1896 patent drawing shows an evolved bipolar coil using tandem chokes to store energy for sudden release into the capacitor, enabling the device to be powered by relatively modest inputs. The chokes are coils often wound on iron cores. They

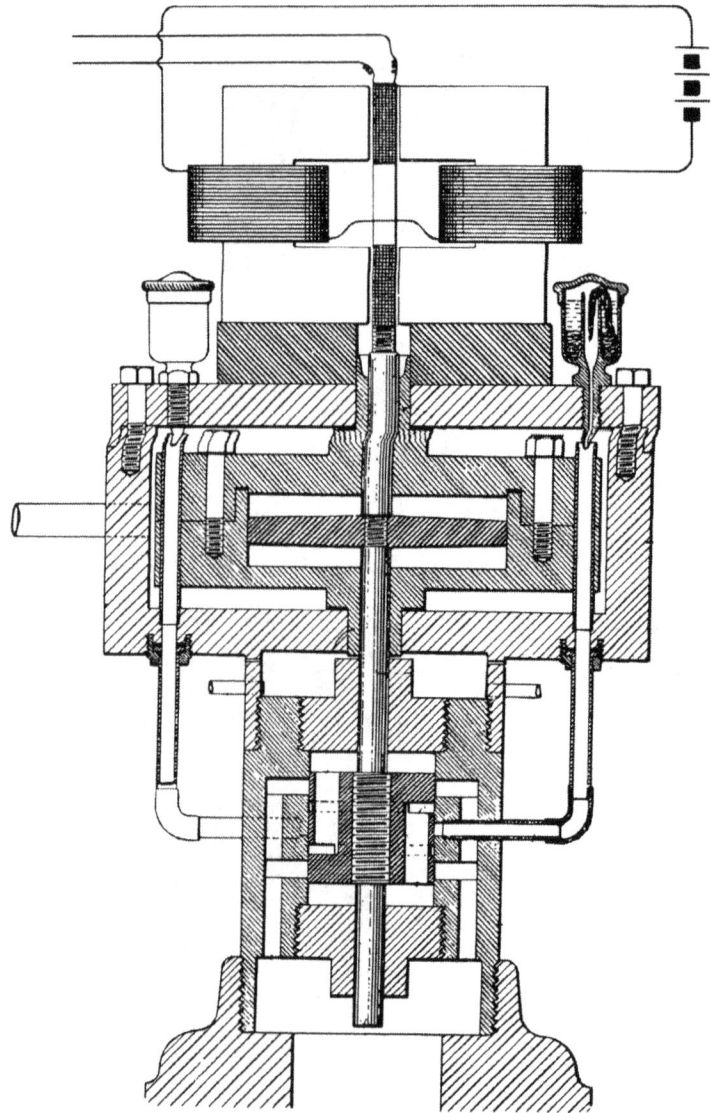

reciprocating engine
From Tesla's Patent No. 511,916

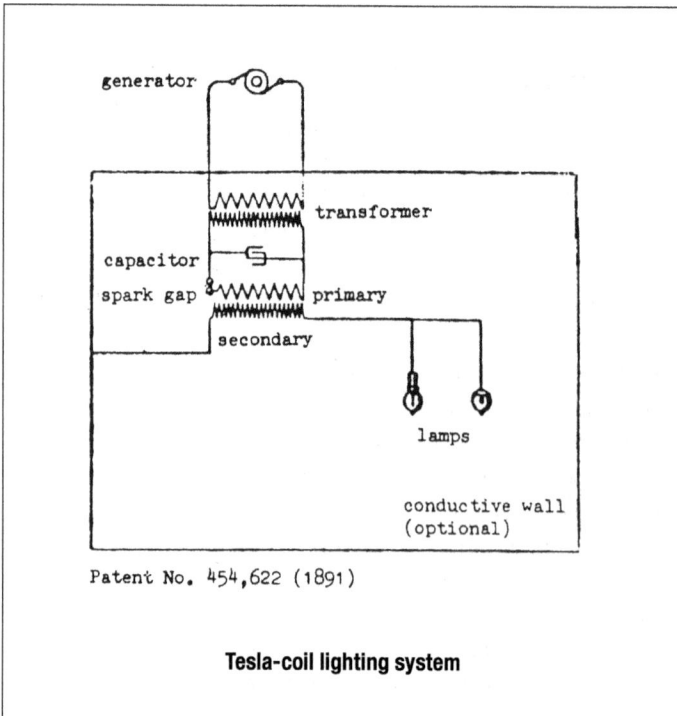

Patent No. 454,622 (1891)

Tesla-coil lighting system

mechanical oscillator by Tesla

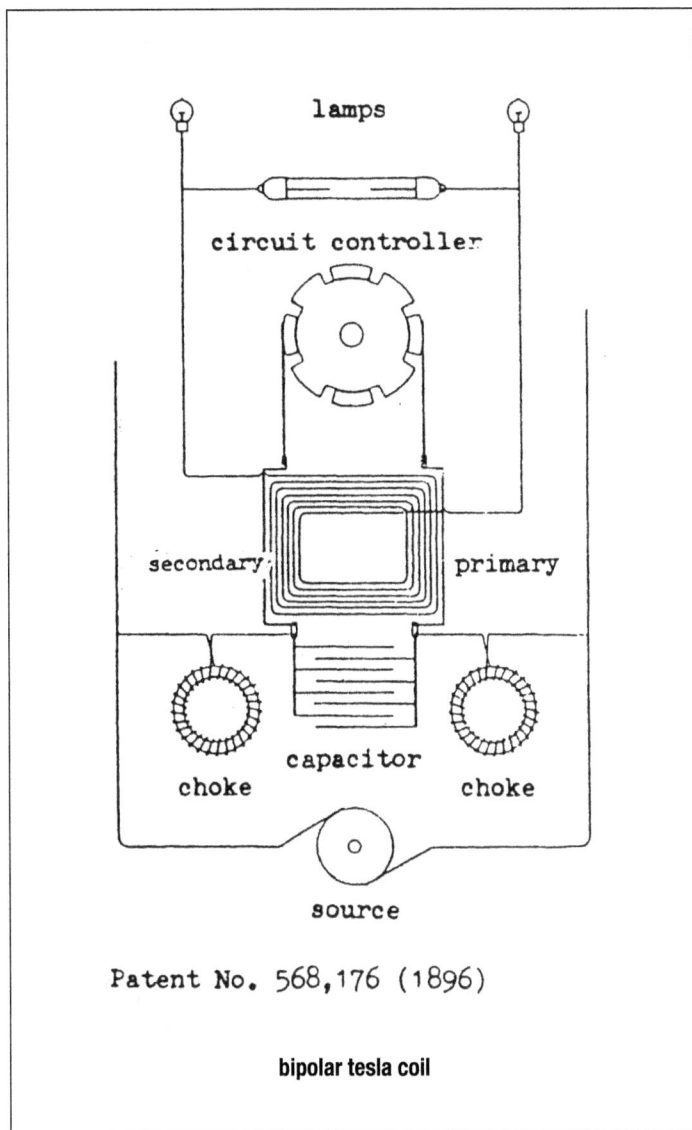

lamps

circuit controller

secondary primary

choke capacitor choke

source

Patent No. 568,176 (1896)

bipolar tesla coil

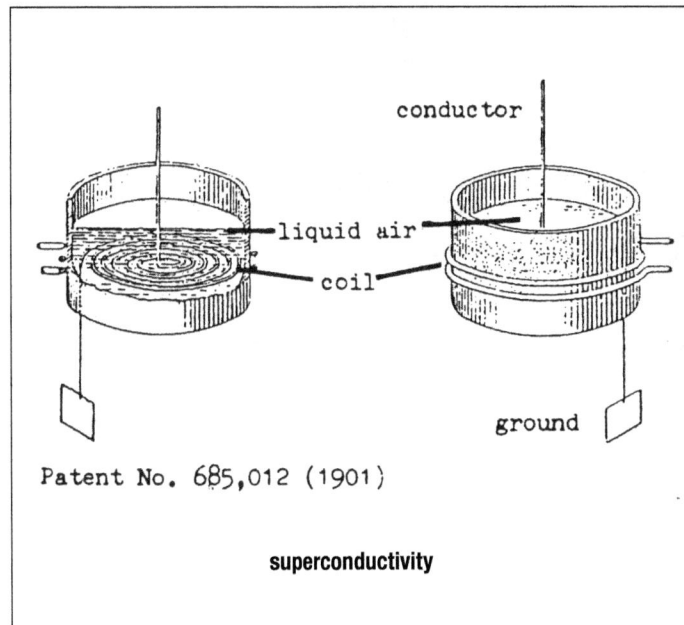

conductor

liquid air

coil

ground

Patent No. 685,012 (1901)

superconductivity

store energy as magnetism. When the current is interrupted, the magnetic field collapses, inducing a current in the coils which rushes in to charge the capacitors.

superconductive

Alternating currents can be sent over long distances with relatively low losses. This is why Tesla's early 60-cycle AC system triumphed over Edison's direct current. The high-frequency, high-potential output of a Tesla coil can travel over relatively light conductors for vastly greater distances than conventional 60 cycle AC. Losses occur to some degree in coronal discharge but hardly at all from ohmic resistance. This type of current also renders conductive materials that are normally nonconductive, rarefied gases for example. You might say that these currents make a medium "superconductive."

It is revealing to reflect upon the unexploited superconductivity of high-frequency, high-voltage Tesla-coil energies these days when science is congratulating itself on new advances in the field. Prior to recent breakthroughs, superconductivity and super-magnetism were low-temperature (cryogenic) phenomena, occurring when circuits were cooled down to near absolute zero. The new superconductivity at less drastically reduced temperatures developed out of the cryogenic work of the last twenty years, and this may be in debt to Tesla, who patented a similar idea way back in 1901. Tesla's patent observes that the deep cooling of conductors with agents like liquid air "results in an extraordinary magnification of the oscillation in the resonating circuit." Imagine the performance of a supercooled Tesla coil.

no electrocution

Since we tend to associate high voltage with possibly fatal electric shock, it may be puzzling to learn that the output of a well-tuned Tesla coil, though in the millions of volts, is harmless. This is customarily thought to be because the amperage is low (it is not necessarily) or it's explained in terms of something called the "skin effect," which means that the current travels over the surface of you instead of through. But the real reason is a matter of neural frequency response. Just as your ears cannot respond to vibrations over about 30,000 cycles, or the eyes to light vibrations above ultra violet, your nervous system cannot be shocked by frequencies over about 2,000 cycles.

Now that you know that these currents are harmless, would you believe they are good for you? Fact is that a whole branch of medicine was founded on the healing effects of certain Tesla-coil frequencies. (See Appendix A.)

CHAPTER 4

Magnifying Transmitter

In 1893 Tesla told a meeting of the National Electric Light Association that he believed it "practical to disturb, by means of powerful machines, the electrostatic conditions of the earth, and thus transmit intelligible signals, and, perhaps power." He said, "It could not require a great amount of energy to produce a disturbance perceptible at a great distance, or even all over the surface of the earth." The ultimate "powerful machine" for these tasks is Tesla's magnifying transmitter.

how it works

The magnifying transmitter has a third or tertiary coil which Tesla called an "extra coil." The extra coil gives the resonant boost of a Tesla-coil secondary but has the advantage of being more independent in its movement. A secondary, being closely slaved to the primary, is inhibited somewhat, its oscillations slightly damped. The extra coil is able to swing more freely. "Extra coils," writes Tesla, "enable the obtainment of practically any emf, the limits being so far remote that I would not hesitate to produce sparks of thousands of feet in this manner."

The engineering challenge of the magnifying transmitter is the containing and proper transmitting of its "immense electrical activities, measured in the tens and even hundreds of thousands of horsepower," as Tesla put it. Containment and effective propagation of this power is the whole point of the design shown, for which Tesla applied for patent in 1902.

The heavy primary is wound on top of the secondary at the base of the tower. The extra coil extends upward through a hooded connection to a conductive cylinder. The terminal is a toroid, a donut-shaped geometry that allows for a maximum of surface area with a comparative minimum of electrical capacity. Since this is a high-frequency device, a relatively low capacity is desirable. To increase the area of the propagating surface, the outside of the toroid is covered with half-spherical metal plates. A subtlety of the design is that the conductive cylinder is of larger radius than the radius of curvature of these plates, since a tighter curve would allow escape of energy. The cylinder is polished to minimize losses through irregularities in the surface. At the center of the top surface sits a pointy plate that serves as a safety valve for overloads so "the powerful discharge may dart out there and lose itself harmlessly in the air."

Tesla advises bringing the power up slowly and carefully so pressure does not build at some point below the terminal, in which case "a ball of fire might break out and destroy the support

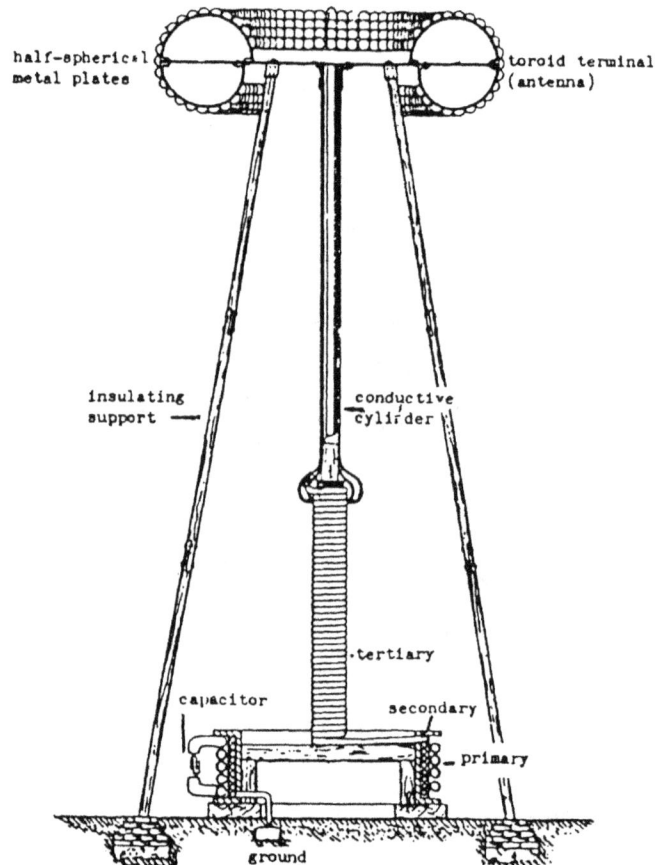

Patent No. 1,119,732 (1902)

or anything else in the way," an event that "may take place with inconceivable violence." Current at the terminal could build to an incredible 4000 amperes.

AC/DC

Wireless power transmission via the magnifying transmitter was the ultimate development of the inventor who had earlier brought alternating-current power to the world with his polyphase system. The predecessor of AC was a direct-current system developed, manufactured, and marketed chiefly by Thomas Edison. Direct current was adequate for serving small areas but was unworkable for long-distance transmission. By contrast, AC could be transmitted for long distances over lighter wires and its voltage could be stepped up for transmission and down for consumption by means of transformers. Tesla invented a new kind of motor (polyphase) that could utilize AC, and he greatly evolved earlier concepts of dynamos to generate AC as well as transformers to step voltage up and down.

Whereas Edison's DC would have been suitable for a society of small, autonomous communities, the evolving system of

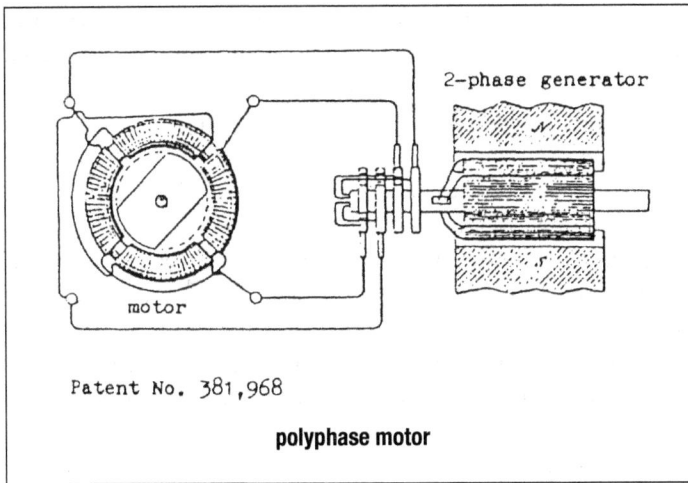

2-phase generator

motor

Patent No. 381,968

polyphase motor

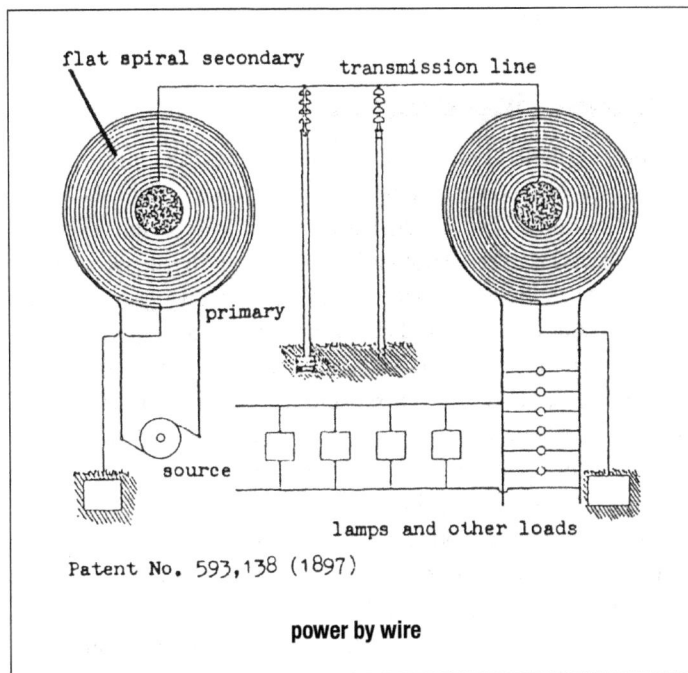

flat spiral secondary

transmission line

primary

source

lamps and other loads

Patent No. 593,138 (1897)

power by wire

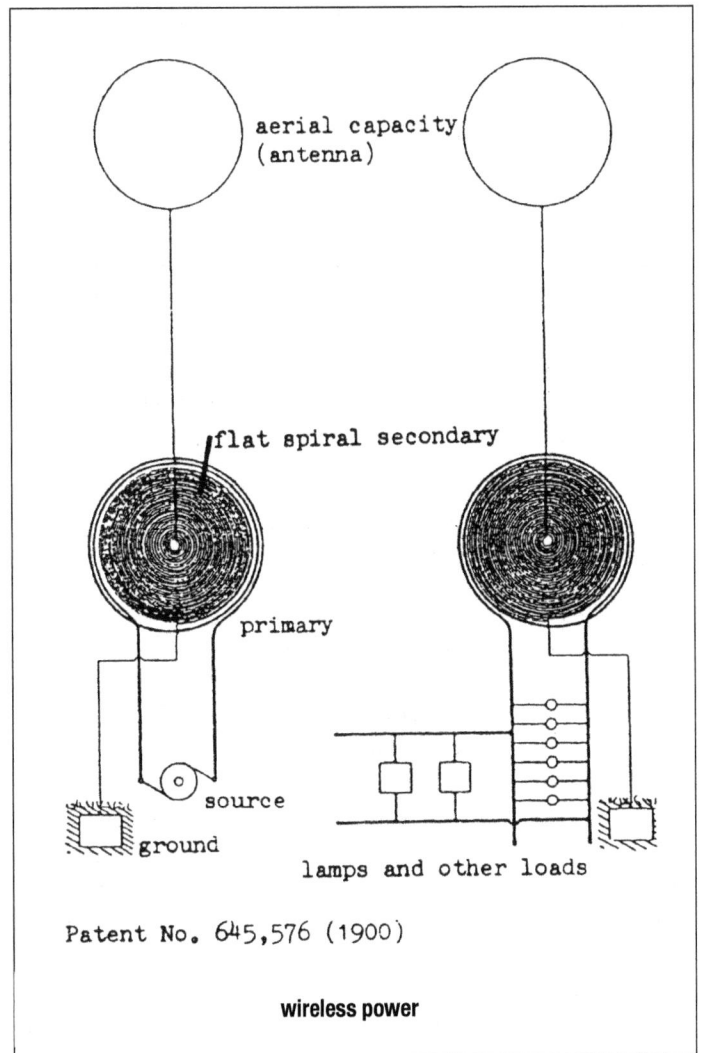

aerial capacity
(antenna)

flat spiral secondary

primary

source

ground

lamps and other loads

Patent No. 645,576 (1900)

wireless power

industrial rule wanted centralized power and needed the long-distance capability of AC to serve huge sprawling populations.

George Westinghouse, an inventor (the airbrake) who, like Edison, turned industrialist (having found that to profit from an invention one must undertake manufacturing and marketing as well) saw the promise in Tesla's polyphase inventions and formed an alliance with the young prodigy.

Westinghouse paid Tesla one million dollars and contracted to pay a royalty of one dollar per horsepower for the polyphase inventions. Later Westinghouse was forced to renege on the royalty. Together, Westinghouse and Tesla triumphed over Edison's DC system and installed the first AC Power facilities, the most notable being the hydro-plant at Niagara Falls. Tesla believed in hydropower. His ultimate energy-magnifying, wireless-power system would have been hydro-based.

The centralized AC electric power system we have today was forced into existence on a colossal scale by utility magnates of Tesla's era, the most prominent being Samuel Insull, who became infamous in some circles for his massive bilking of the investing public and famous in others for hammering together the electric power complex now in place. This complex has developed into a federally protected monopoly with greater capital wealth than any other industry in the U.S. In the order of energy sources used, Tesla's hydropower has been left well behind by the burning of coal and oil fuels and even by the nukes in kilowatt hours produced. So went another Tesla dream.

Tesla was a celebrity in his polyphase heyday, but today his celebrity is as an underground cult figure known for his radically progressive energy-magnifying, free-energy, and wireless power inventions, which, of course, have no place in the established system.

power by wire

Prior to his wireless power inventions, Tesla patented in 1897 a high-frequency system that transmitted power by wire. The system used previously unheard of levels of electric potential. He notes that at these voltages, conventional power would destroy the equipment, but that his system not only contains this energy but is harmless to handle while in use. This system is not a circuit in the usual sense but a single wire without return. This is a fundamental concept in Tesla technology that enables radio and wireless power.

The system shown employs the familiar Tesla-coil configurations at both sending and receiving ends. The primary circuit (power source, capacitor, spark gap) is represented in the drawing by the generator symbol. The secondary coil is a flat spiral. An advantage in this coil design is that the voltage adjacent to the primary, where arcing across could occur, is at zero and soars to high values as the coil spirals inward. The same patent also shows a cone-shaped secondary in which the primary is at the base of the cone, which is at zero potential.

wireless power

The drawing for Tesla's wireless power patent looks like the earlier power-by-wire patent except now spherical antenna terminals replace the transmission lines, which are dropped out of the picture almost as if they were redundant The ball aerial is peculiar to Tesla, as is the toroid, and you wonder why nothing like them have appeared since.

In this 1900 patent, wireless power is not represented as an earth-resonant system. Here Tesla talks about transmission through "elevated strata." The patent contains much discussion of how rarefied gases in the upper atmosphere became quite conductive when there is applied "many hundred thousand or millions of volts." Balloons are suggested to send the antennas aloft.

Appreciate that Tesla in this patent has invented nothing less than the principles of radio. Tesla recognizes only a quantitative difference between sending radio signals and broadcasting electric power. Both involve sending and receiving stations tuned to one another by means of Tesla-coil circuits.

Tesla's wireless power would be the ultimate centralized electric system, an orthocratic dream, but for the fact that the technology is too simple. Reception of power could be achieved just by raising an antenna terminal, planting a ground, and connecting simple Tesla-coil circuitry in between. Although Tesla himself patented a couple of electric meters for high frequencies, it would be all too easy for consumers to tune in for free, just as many today bootleg pay TV signals using illicit equipment far more sophisticated. It is no wonder, then, that the electric power establishment didn't welcome this invention. This was one problem. Another was that the established electric power system would have to be relegated to another great pile of scrap, and maybe an important piece of the established system of political power as well. Tesla's announced dream was to use hydro sources where available and through wireless power broadcast that energy around the planet.

Such a scheme would not be readily embraced by powers that sustain their rule by keeping populations poor and weak. Centralized control of energy, as well as other resources, is, of course, believed to be essential to civilized rule, at least as far as thinking on that subject has progressed in this era. Moreover, no multinational political system was in existence, or is now for that matter, that could implement a technology of such global implications.

Tesla was blind to such considerations. His commitment, his overriding priority as a technological purist, was to take machine possibilities to their logical conclusions. Today, if wireless power were seriously proposed, there would no doubt be at least one political problem that would not have arisen in Tesla's time: resistance from environmentalists. What would an environmental impact report have to say about biologic hazards? A Navy submarine communication system that uses extremely low frequency (ELF) waves, below 10 cycles per second, has been challenged by environmentalists, as have microwave and 60-cycle high-voltage transmission lines.

engineering details

Patents normally don't give many quantitative specifics, but Tesla's wireless power patent does give some detail about the prototype he used to conduct a demonstration before skeptical patent examiners. A 50,000-volt transformer charged a capacitor of .004 mfd, which discharged through a rotary gap that gave 5,000 breaks per second. The eight-foot diameter primary had just one turn of stout stranded cable. The secondary was 50 turns of heavily insulated No. 8 wire wound as a flat spiral. It vibrated at 230-250 kilocycles and produced two to four million volts.

This coil evolved into the huge experimental magnifying transmitter Tesla describes in his Colorado Springs notes. Housed in a specially built lab 110 feet square, the device used a 50,000 volt Westinghouse transformer to charge a capacitor that consisted of a galvanized tub full of salt water as an electrolyte, into which he placed large glass bottles, themselves containing salt water. The salt water in the tub was one "plate" of this capacitor, the salt water inside the bottles the other "plate," and the bottle glass was the dielectric. Various capacities were tried, incremental changes being made by connecting more or fewer bottles. A variable tuning coil of 20 turns was connected to the primary which consisted of two turns of heavy insulated cable that ran around the base of the huge fence-like wooden framework. The secondary had 24 turns of No. 8 wire on a diameter of 51 feet.

Various extra coils were tried, the final version being 12 feet high, 8 feet in diameter, and having 100 turns of number-8 wire. The terminal was a 30-inch metal ball, rubber-coated, adjustable for height on a 142-foot mast. The huge transmitter could vibrate from 45 to 150 kilocycles. Even with the big transformer, this bill of materials does not seem inaccessible to enterprising people, and the technology does not seem so abstruse, so it is no wonder that people have gotten together to build magnifying transmitters and experiment with wireless power without support from corporations or government.

earth resonance

Among the appealing features of Colorado Springs for Tesla was the region's frequent and sensational electrical storms. For Tesla, lightning was a joyous phenomenon. Biographers report that, during storms Tesla would throw open the windows of his New York lab and recline on a couch for the duration, muttering to himself ecstatically. In Colorado Springs he tuned in and tracked lightning storms using rudimentary radio receiving equipment. He thereby determined that lightning was a vibratory phenomenon which set up standing waves bouncing within the earth at a frequency resonantly compatible with the earth's electrical capacity. This earth-resonant frequency, he reasoned, was the ideal frequency for wireless power transmission, and he tuned his ultimate magnifying transmitter accordingly.

The literature contains various reports on exactly what this frequency is. Some say 150 kilocycles, which would be at the upper range of the Colorado Springs transmitter. Others give frequencies considerably lower, 11.78 cycles, 6.8 cycles, frequencies Tesla's transmitter may have achieved harmonically.

With reinforcement from the earth resonance, the power would actually increase in the process of transmission.

In one memorable experiment with the Colorado Springs transmitter, Tesla shot from the aerial ball veritable lightning bolts of 135 feet, producing thunder heard 15 miles distant, and in the process burned out the municipal generator. In another experiment he lit up wirelessly, at a distance of 26 miles from the lab, a bank of 10,000 watts worth of incandescent bulbs.

Two years after Colorado Springs, Tesla applied for patent for the far more refined magnifying transmitter shown at the opening of this chapter, a patent that was not granted until a dozen years later. In this patent he no longer speaks of energy broadcast through the "upper strata" of the atmosphere but of a "grounded resonant circuit."

Tesla predicted that his magnifying transmitter would "prove most important and valuable to future generations," that it would bring about an "industrial revolution" and make possible great "humanitarian achievements." Instead, as we shall see in Part 2, Radio Tesla, the magnifying transmitter would become Tesla's Waterloo.

CHAPTER 5

Lighting

In 1891 Tesla said that existing methods of lighting were "very wasteful," that "some better methods must be invented, some more perfect apparatus devised." Tesla went and did just that. Edison's enduring filament bulb operates with six percent efficiency, the rest going off as heat. The filament, which cooks at about 4,000 degrees, eventually breaks without warning. Primarily it's all the on-and-off cycles that kills it. However, the design must also be held accountable.

I prefer the filament bulb to the new bulbs for its warm quality for room lighting, and because I resent the AM-radio interference generated by the florescent, including the compact fluorescent. The new lighting technology also gives us the LED cluster, but this is a heater too and must be heavily heat-sinked. I use the filament bulb almost exclusively.

My bulbs last forever. Why? I run them on variable-voltage transformers. Tesla-coilers call these "variacs." I operate the bulbs at voltages well below the 120-volt spec, rarely over 85 or 90. For a burst of illumination, I can run the variac in overdrive up to 140 volts, but normally I tax my filaments with only about 35 to 90 volts. And they last "forever." What does this say about the engineering of the conventional 120-volt filament bulb: built to fall apart?

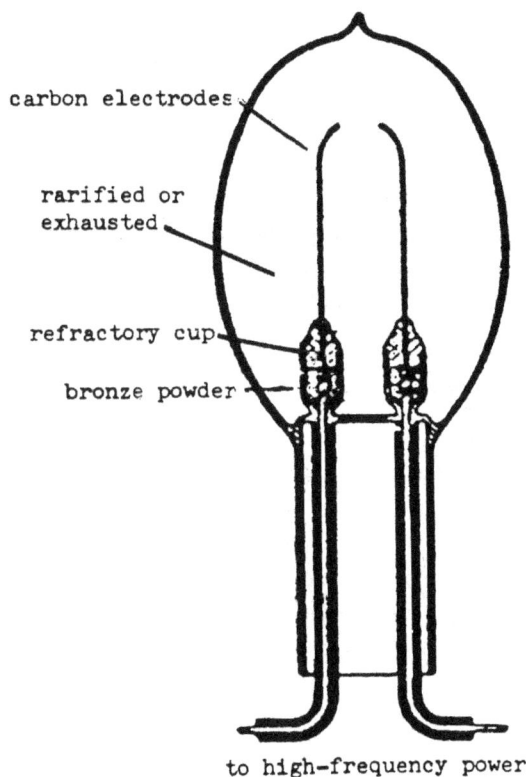

carbon electrodes

rarified or exhausted

refractory cup

bronze powder

to high-frequency power

Patent No. 455,069 (1891)

Tesla, who invented neon lighting may have inspired the technology that brings us the fluorescent. Today's version is no model of efficiency either. Its inner surfaces are stimulated to phosphorescence by energy-consuming filament-like cathodes that also burn out, and the lit-up tube would present a dead short to the current if it were not for the so-called "ballast transformer," an inductance placed in the circuit to oppose and thus eat up yet more current.

What sent Tesla into an exploration of high frequency phenomena was his conviction that these rapid vibrations held the key to a superior mode of lighting. These explorations were not Tesla's first venture into lighting. His very First U. S. patent (1885) is for an improvement in the arc lamp. He used an electromagnet to feed carbons to the arc at a uniform rate to produce a steadier light (No. 335,785). Early arc lamps produced a brilliant blue-white light, good for street lighting but not for the home, and they emitted noxious fumes. Home lighting was by gas.

Street arc-lighting used series circuits. Edison introduced the parallel circuit, and designed his lamp for such a circuit. Edison wanted to be first in the electric business and announced to the press that he had an operable bulb before he actually had one that worked. Edison introduced the big-scale production and sale of electric power itself on the model of gas lighting, a major industry at the time. (Early electric installations used the plumbing of gas lighting for wire conduit.) Gas-lighting systems set the model for central energy production, for mass distribution, for billing, and ultimately for the electric grid. Centralization and mass-distribution may be necessary for a natural gas system, but electric power is most efficiently produced on location in home or factory, contrary to the centralized orthocratic control model in place. The old model is now being extended to huge centralized solar and wind farms, distant from users and scars upon the remote landscapes they occupy.

When Tesla's AC system was established, it was grafted onto Edison's DC system, greatly extending its range and efficiency, but it was still the Edison centralized mass-distribution system with parallel circuits and incandescent lighting.

a better way

Tesla patented both his spark-gap oscillator and his Tesla coil specifically as power sources for a new lighting system that used currents of high frequency and high potential. Lest you get the impression that a lone genius named Tesla invented this new form of lighting out of the blue, you should know that others before him had used high frequencies to stimulate light, and others, like Sir William Crookes, had done the same with high potentials, but Tesla was the first on record to put the two together.

In Jules Verne's 1872, novel *A Journey to the Center of the Earth*, the narrator tells of a brilliant portable battery lamp used by the underground explorers. It was powered by a Ruhmkorf coil, a high voltage buzzer-type induction coil popular among early

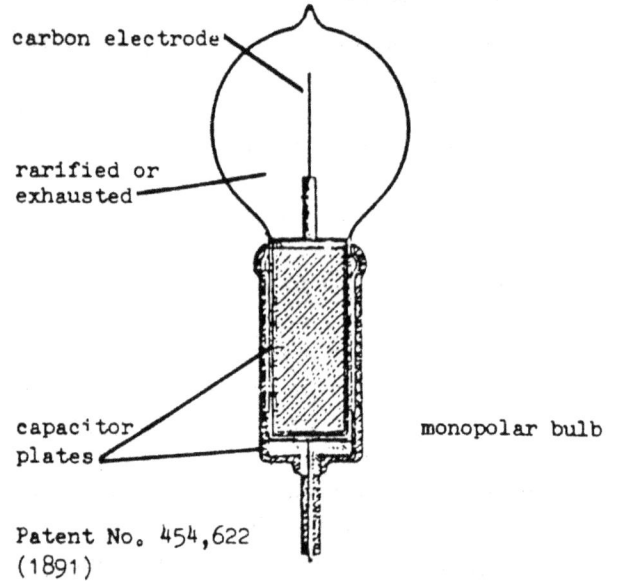

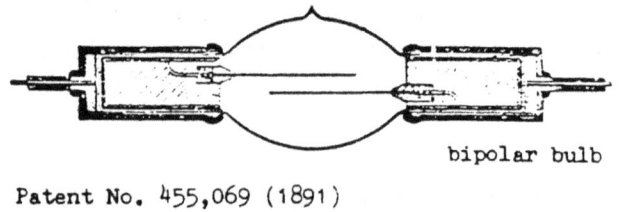

carbon electrode

rarified or exhausted

capacitor plates

monopolar bulb

Patent No. 454,622 (1891)

bipolar bulb

Patent No. 455,069 (1891)

capacitor bulbs

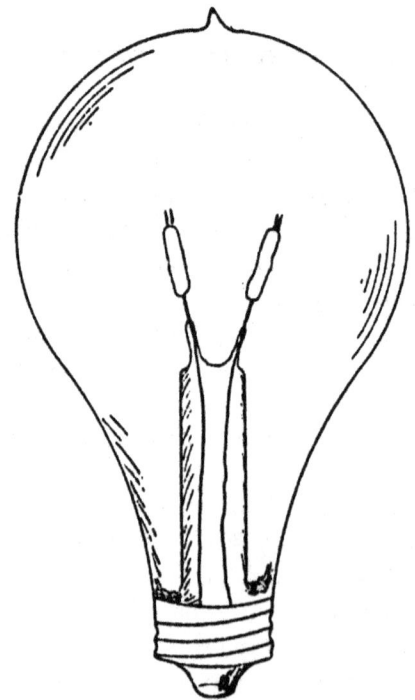

solid bodies become incandescent

"pure light"

Tesla invented a variety of lamps, not all of which show up in his patents. He lit up solid bodies like carbon rods set inside vacuum bulbs, or solid bodies set inside bulbs containing various inert gases at low pressure (rarefied). He noted that "tubes devoid of any electrodes may be used, and there is no difficulty in producing by their means light to read by." But he noted that the effect is "considerably increased by the use of phosphorescent bodies, such as yitria, uranium glass, etc." Here Tesla lays the foundation for fluorescent lighting.

Applied to such lamps were currents at potentials ranging from a lower limit of 20,000 volts up to voltages in the millions and vibrations of 15,000 cycles per second and up. Tesla dreamed of creating what he called "pure light" or "cold light" by generating electric vibrations at frequencies that equaled those of visible light itself. Light produced by this direct and efficient means would require vibrations of 350 to 750 billion cycles, but Tesla believed such oscillations, far above those attainable by his coils, would someday be achieved. Even so, his rarefied gas-tube lamps produced a light that more closely approximated natural daylight than any other artificial source. Tesla's light is like the "full-spectrum" light which is coming to be recognized as far more healthful than Edison's incandescent and particularly more healthful than conventional fluorescent, and is believed by some health practitioners actually to have healing properties.

no sudden burn-out

Tesla's gas tube lamps burn indefinitely, as do today's neon tubes, for there is nothing within to be consumed. Tesla's lamps that contain electrodes like carbon rods, however, do undergo some deterioration. In Tesla's words, "a very slow destruction and gradual diminution in size always occurs, as in incandescent filaments; but there is no possibility of sudden and premature disabling which occurs in the latter by the breaking of the filament, especially when incandescent bodies are in the shape of blocks." In vacuum lamps, the life of the bulb depends upon the degree of exhaustion, which can never be made perfect.

The electrodes within bulbs may glow at high temperatures, raising the problem of how to conduct energy to them, since wires or other metallic elements will melt. The problem must be addressed in lamp design. For example, in the incandescent lamp shown at the opening of this chapter, the lead-in wires connect to the hot electrodes via bronze powder contained in a refractory cup. Tesla may have designed his capacitor-base bulbs to help address this same problem.

high heat

Tesla's search for the ideal electrode is reminiscent of Edison's search for the long-lasting filament: "The production of a small electrode capable of withstanding enormous temperatures," said

electrical experimenters. The Ruhmkorf coil stimulated a lamp (type unspecified but probably a gas tube) which produced "the light of an artificial day." The lamp had such a low current draw that the battery lasted throughout the subterranean adventure. Verne evidently was drawing upon the experimental knowledge of his day for what he calls "this ingenious application of electricity to practical purposes."

Modern neon lighting is high potential at 2,000 to 15,000 volts. Neon sign transformers are good for powering Tesla coils, but are low-frequency (60-cycle) high-voltage devices. The compact fluorescent is a high-frequency device, but not high-voltage. Neon is high-voltage, and so is 7,500-volt "cold cathode" fluorescent used in industrial lighting.

Circa 1900 Tesla experimented with luminous tubes bent into alphabetic characters and other shapes. Although much of today's neon is still simplistic Tesla, being driven by 60-cycle high-voltage transformer power alone without the benefits of high-frequency excitation, it should suggest to us the amazing efficiency of high-potential lighting, since a single 15,000-volt neon transformer drawing only 230 watts can light up a tube extending up to 120 feet. The new electronic neon transformer is a 3-kilocycle device (page 94).

How superior is the economy of Tesla high-potential, high-frequency lighting over Edison's incandescent? Tesla says "certainly 20 times, if not more" light is obtained for the same expenditure of energy.

Tesla, "I regard as the greatest importance in the manufacture of light." One of the electrodes he tried was a small "button" of carbon which he placed in a near-vacuum. Tesla regarded the high incandescence of the button to be a "necessary evil." For lighting purposes, It was the incandescence of the gas remaining in the mostly evacuated chamber that was important.

But the carbon-button lamp proved to have some remarkable properties beyond its use for illumination. When the voltage was turned up, the lamp produced such tremendous heat that the carbon button rapidly vaporized. Tesla experimented extensively with this fascinating phenomenon. For the button of carbon he substituted zirconia, the most refractory substance available at the time. It fused instantly. Even rubies vaporized. Diamonds, and, to a greater degree, carborundum, endured the best, but these could also be vaporized at high potentials.

space heating

Did Tesla work on the problem of electric space heating. I have read that he contributed to the development of a high-frequency induction heating. Certainly the huge current draw of conventional electric heaters which use resistive elements argues for some inventiveness in this area. Tesla did observe that the discharges from a Tesla coil resembled "flames escaping under pressure" and were indeed hot. He reflected that a similar process must take place in the ordinary flame, that this might be an electric phenomenon. He said that electric discharges might be "a possible way of producing by other than chemical means a veritable flame which would give light and heat without material consumed."

The behavior of the carbon-button lamp suggests that a new heating mode might be found in the effects of high-frequency currents in a vacuum.

lighting up the sky

Hold a fluorescent tube near a Tesla coil and it will light up in your hand. This is true of any tube or bulb with vacuum or rarefied gas. A more efficient way is to ground one end of the tube and to put a length of wire as a sort of antenna on the other. Better yet, put a coil of wire that resonates with the secondary in series with the tube and ground and you have the optimal wireless power arrangement. Tesla conducted many experiments with different arrangements like this, using on some occasions the widely available Edison filament incandescent, which lighted up more brilliantly than usual because of the effects of high frequencies on the bulb's rarefied interior.

Inside his New York lab, Tesla strung a wire connected to a Tesla coil around the perimeter of the room. Wherever he needed light he hung a gas tube in the vicinity of this high-frequency conductor.

Tesla had a bold fantasy whereby he would use the principle of gas luminescence to light up the sky at night. High frequency electric energy would be transmitted, perhaps by an ionizing

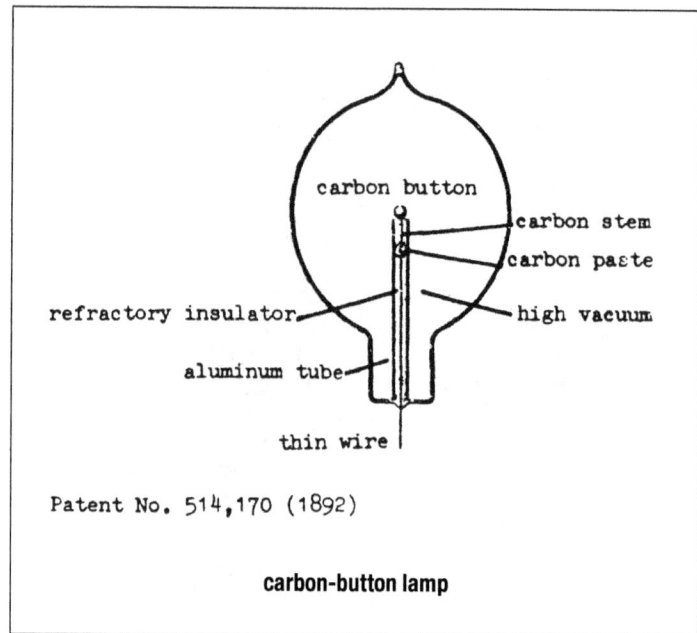

Patent No. 514,170 (1892)

carbon-button lamp

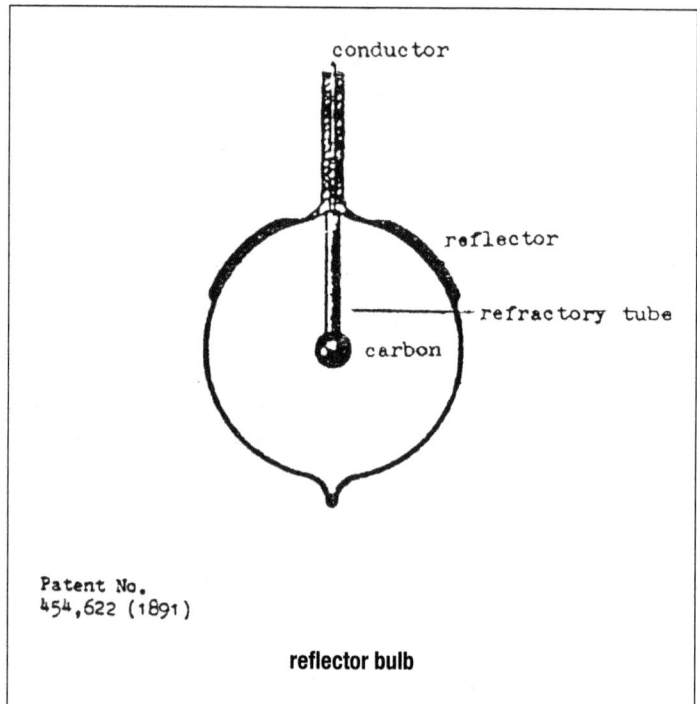

Patent No. 454,622 (1891)

reflector bulb

beam of ultraviolet radiation, into the upper atmosphere, where gases are at relatively low pressure, so that this layer would behave like a luminous tube. Sky-lighting, he said, would reduce the need for street lighting, and facilitate the movement of ocean-going vessels. The Aurora Borealis is an electrical phenomenon that works on this principle, the effects of cosmic eruptions such as those from the sun being the source of electric stimulation. I, for one, am grateful that this particular Tesla fantasy never materialized since it is difficult enough to see the stars with existing light pollution, and there might be undesirable biological impacts as well.

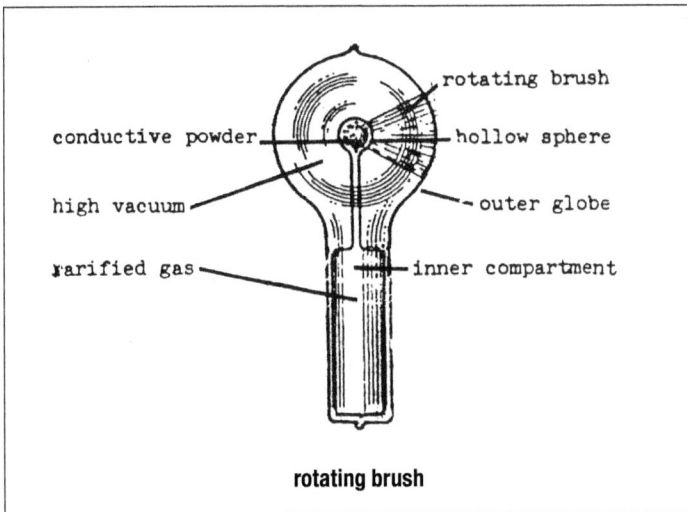

rotating brush

rotating " brush"

Tesla took an evacuated incandescent type lamp globe, suspended within it at dead center a conductive element, stimulated that element with high-voltage currents from an induction coil, and thus created a beam-like emanation, a "brush" discharge that was so eerily sensitive to disturbances in its environs that it seemed to be endowed with an intelligent life of its own. The device works best if there is no lead-in wire. In the bulb shown, every measure has been taken to construct it so it is free from its own electrical influence. The bulb could be stimulated capacitively by applying energy to metal foil wrapped around its neck.

Thus excited, "an intense phosphorescence then spreads at first over the globe, but soon gives place to a white misty light," observes Tesla. The glow then resolves into a directional "brush" or beam that will spin around the central element. So responsive is the brush to any electrostatic or magnetic changes in its vicinity that "the approach of an observer at a few paces from the bulb will cause the brush to fly to the opposite side." A small, inch-wide permanent magnet "will affect it visibly at a distance of two meters, slowing down or accelerating the rotation according to how it is held relatively to the brush."

Tesla never patented the rotating brush or used it in any practical application, but he believed it could have practical applications. He saw one use in radio where the device could conceivably be adapted to being a most sensitive detector of disturbances in the medium. The rotating brush appears to be a precursor of the plasma-globe toys, which are sometimes called "Tesla globes." Tesla's rotating brush certainly numbers among his lost inventions and deserves further research. One can imagine uses in an advanced, non-digital artificial intelligence, in radio detection, and even in a television on a new principle.

Tesla's new lighting was famous in its time. Tesla, the promoter, saw to it. He conducted demonstrations at lectures before the electric industry associations, before large audiences in rented halls, and before select groups of influential New Yorkers in his lab. His articles about the new lighting were published in the popular scientific press and reported in the newspapers. Still, it did not catch on with the powers-that-be who no doubt saw in it Tesla's perennial pile-of-scrap problem.

But, I wonder, would the whole electric distribution system have to be scrapped to implement the efficiencies of Tesla lighting? Conceivably, the new lighting could be run off of local oscillators at the consumer end, the old power distribution system remaining intact, which is roughly what we have now with the compact fluorescent.

CHAPTER 6

Transport

Tesla speculated, "Perhaps the most valuable application of wireless energy, will be the propulsion of the flying machine, which will carry no fuel and be free from any limitations of the present airplanes and dirigibles." The possibility of electric flight intrigued Tesla, though he never did patent an electric aircraft. But he did patent an electric railway using his high-frequency, high-potential electricity in a by-wire mode, and also patented a radical aircraft that, while not electric, did have an advanced power plant: his disk turbine.

Tesla's railway and aircraft can be numbered among the lost inventions. The closest transport technology has come to putting any of Tesla into actual practice is with diesel-electric power

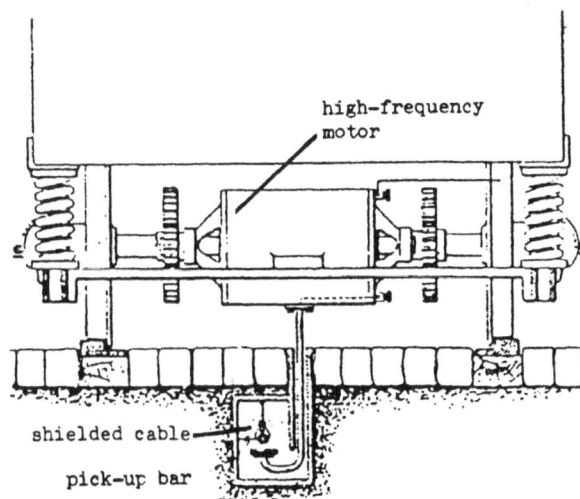

Patent No. 514,972 (1892)

high-frequency railroad

using Tesla polyphase motors, an early and notable example of which was the ocean liner Normandy. In the field of transport Tesla is more commonly identified with antigravity flight and with UFOs. Although this identification is based upon nothing more than a few public utterances, his suggestions charge the imagination with possibilities.

high-frequency railway

Tesla's high-frequency, high-potential railway picks up its power inductively without the use of the rolling or sliding contacts used in conventional trolley or third-rail systems. A pick-up bar travels near a cable carrying the oscillating energy. This cable, which Tesla specifically invented to carry such currents, is the precursor of the grounded shielded cable used today to carry TV and other high-frequency signals. But unlike today's cables, which carry energy only of signal strength and shield by means of a continuous grounded static screen of fine braided copper wire, Tesla's high-voltage cable uses metal pipe or screen that is broken up into short lengths, "very much shorter," says Tesla in his patent, "than the wave lengths of the current used." This feature reduces loss. Since the shielding must not be interrupted, the short sections are made to overlap but are insulated from one another. To further reduce loss to ground, an inductance of high ohmic resistance or a small capacity is placed in the ground line.

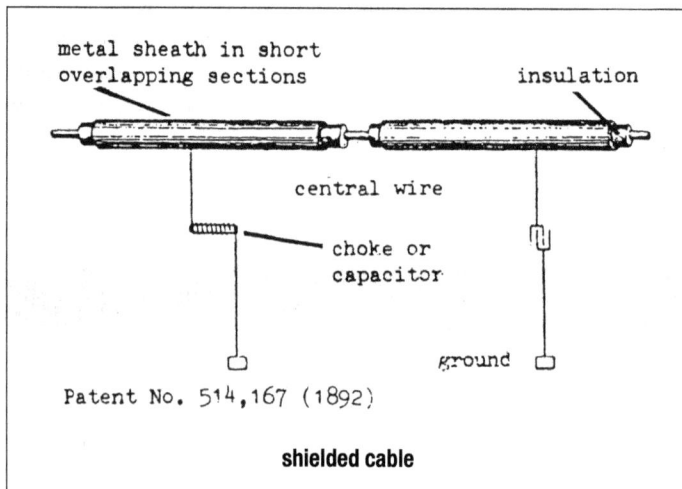

Patent No. 514,167 (1892)

shielded cable

motor mystery

A conundrum raised by Tesla's railway patent is that the vehicle is powered by an electric motor, but nowhere among Tesla's inventions is to be found an electric motor run by high-frequency currents. Was Tesla planning to use a lower frequency here, something under 1,000 cycles? Did he have a converter in mind that could bring the frequency down? Or did Tesla invent a high-frequency motor that never made it into patent, an invention that may be among his unpublished notes? Anyway, Tesla proceeds in many of his discussions of high-frequency power as if this

problem were solved. I've seen references to the existence of such a motor. Free-energy inventor, Hermann Plauson, (next chapter) refers to high-frequency motors. These motors have magnetic cores made of very thin laminations insulated from each other, a design that would limit damping effects.

turbine aircraft

Tesla's only patented aircraft is a vertical take-off and landing (VTOL) plane that he intended as an improvement upon the helicopter, already invented at this time (1921). "The helicopter type of flying machine, especially with large inclination angle of the propeller axis to the horizontal, at which it is generally expected to operate, is quite unsuitable for speedy aerial transport; it is incapable of proceeding horizontally along a straight line under prevailing air conditions; it is subject to dangerous plunges and oscillations and it is almost certainly doomed to destruction in

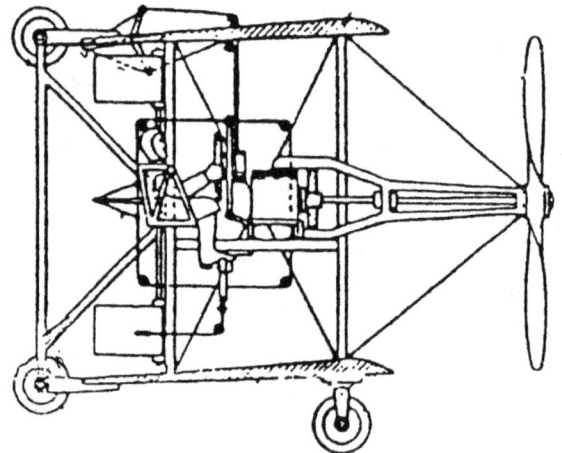

VTOL aircraft

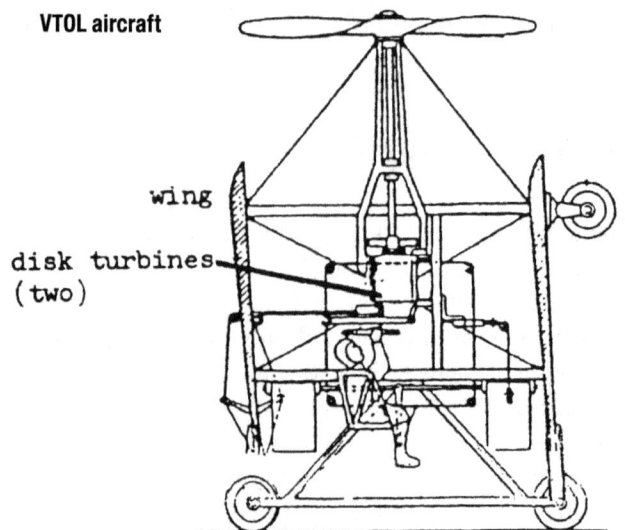

Patent No. 1,655,113 (1921)

case the motive power gives out." Advances in helicopter design may have mitigated some of these problems, but at least the last still holds true.

Tesla's craft, which has a large wing area, is powered by two disk turbines. The engineering problem of swinging the pilot and passengers around 90 degrees after take-off is solved at least to Tesla's satisfaction.

electric flight

Tesla's dream electric aircraft would be powered by earth-based magnifying transmitters. "Aerial machines will be propelled around the earth without a stop." Also, in 1900, he predicted a "cold coal" battery with such output that "a practical flying machine" would be possible. Such a battery also "would enormously enhance the introduction of the automobile."

Tesla fantasized a personal "aerial taxi" which could be folded into a six-foot cube and would weigh under 250 lbs: "It can be run through the streets and put in a garage, if desired, just like an automobile" Explaining how his earth-resonant wireless-power system could energize vehicles aloft, he said, "power can be readily supplied without ground connection, for, although the flow is confined to earth, an electromagnetic field is created in the atmosphere surrounding it."

Tesla believed such a system to be the ultimate method of man-made flight: "With an industrial plant of great capacity, sufficient power can be derived in this manner to propel any kind of aerial machine. This I have always considered the best and permanent solution to flight. No fuel of any kind will be required as the propulsion will be accomplished by light electric motors operated at great speed."

antigravity

Tesla wrote in 1900 of an antigravity motor: "imagine a disk of some homogeneous material turned perfectly true and arranged to turn in frictionless bearings on a horizontal shaft above the ground. Now, it is possible that we may learn how to make such a disk rotate continuously and perform work by the force of gravity." To do so, he said, "we have only to invent a screen against this force. By such a screen we could prevent this force from acting on one-half of the disk, and rotation of the latter would follow. "

Does it not follow then, that such a gravity screen could also be used to levitate a vehicle? Tesla held no patent on such a device, but recently a Tesla antigravity coil has surfaced (see Teslapress). Tesla inevitably pops up in the literature of antigravity and UFOs. This may be because Tesla was a prominent exponent of a physics in which antigravity seems more feasible because gravity is better explained.

In 1931 the editor of *Science & Mechanics*, Hugo Gemsback reported, "It is believed by many scientists today that the force of gravitation is merely another manifestation of electromagnetic waves."

Edward Farrow, a New York inventor, reported in 1911 an antigravity effect produced by a ring of spark gaps. When the gaps were fired, the device, called a "condensing dynamo," lost one-sixth of its weight.

T. Henry Moray wrote that "Frequencies may be developed which will balance the force of gravity to a point of neutralization." Antigravity researcher Richard Lefors Clark places the frequency of gravity's vibrations right at "Nature's neutral center in the radiant energy spectrum," above radar and below infrared, at 10 to the 12th cycles per second.

CHAPTER 7

Free-Energy Receiver

For starters, think of this as a solar-electric panel. Tesla's invention is very different, but the closest thing to it in conventional technology is in photovoltaics. One radical difference is that conventional solar-electric panels consist of a substrate coated with crystalline silicon; the latest use amorphous silicon. Conventional solar panels are expensive, and, whatever the coating, they are manufactured by esoteric processes. But Tesla's "solar panel" is just a shiny metal plate with a transparent coating of some highly insulating material which today could be many layers of a spray plastic.

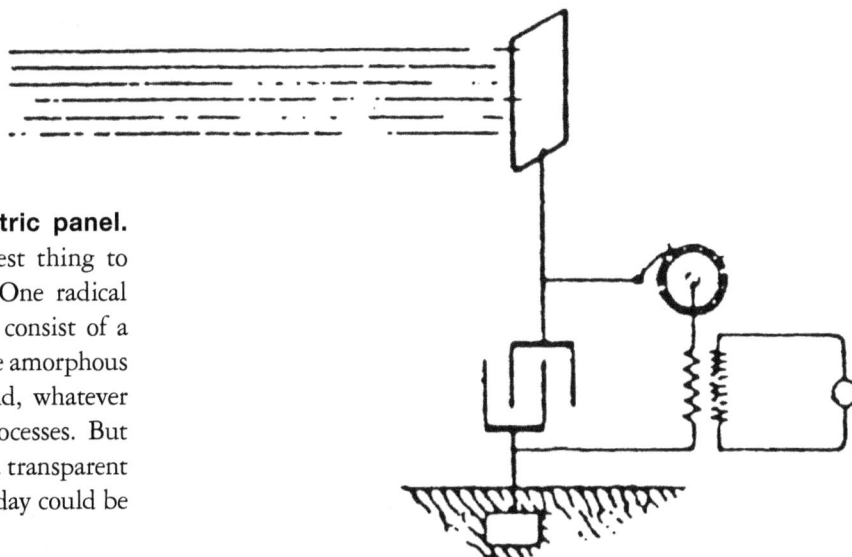

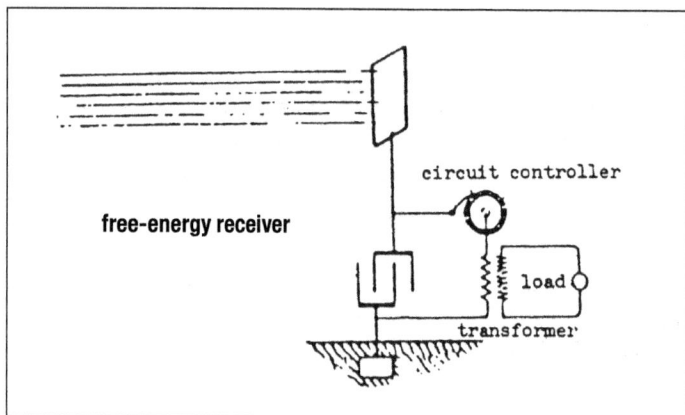

free-energy receiver

circuit controller

load

transformer

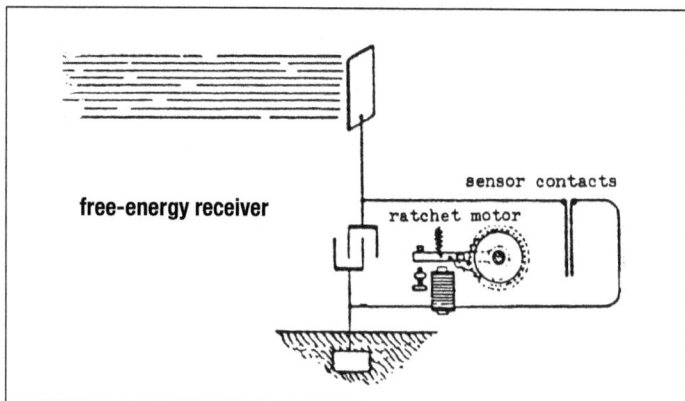

free-energy receiver

sensor contacts

ratchet motor

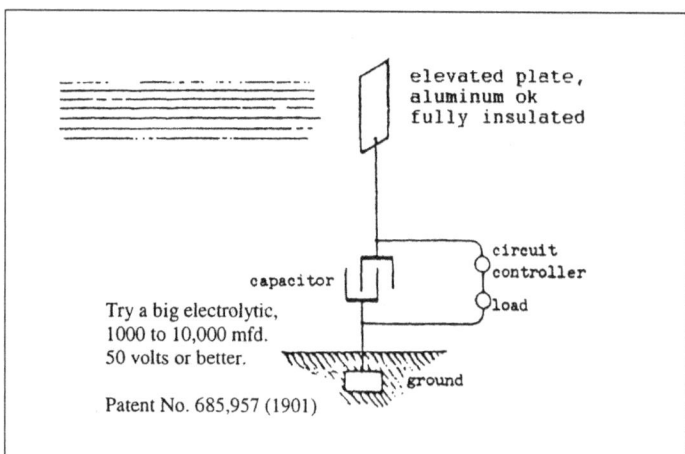

elevated plate,
aluminum ok
fully insulated

capacitor

circuit
controller

load

Try a big electrolytic,
1000 to 10,000 mfd.
50 volts or better.

Patent No. 685,957 (1901)

ground

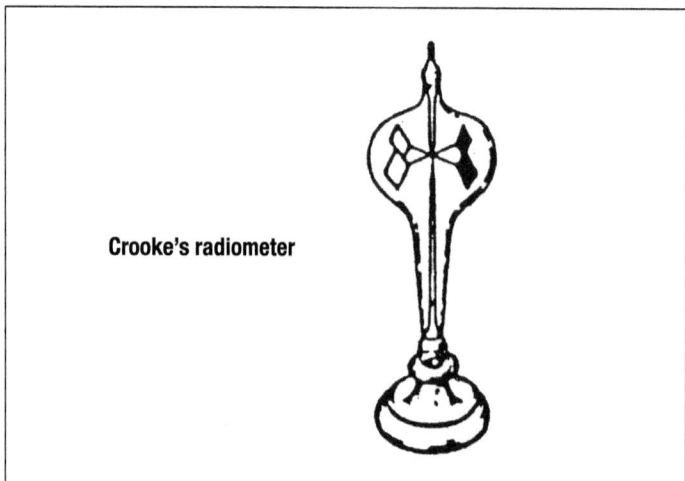

Crooke's radiometer

Stick one of these antenna-like panels up in the air, the higher the better, and wire it to one side of a capacitor, the other going to a good earth ground. Now the energy from the sun is charging that capacitor. Connect across the capacitor some sort of switching device so that it can be discharged at rhythmic intervals, and you have an electric output. Tesla's patent is telling us that it is that simple to get electric energy. The bigger the area of the insulated plate, the more energy you get. But this is more than a "solar panel" because it does not necessarily need sunshine to operate. It also produces power at night.

Of course, this is impossible according to official science. For this reason, you could not get a patent on such an invention today. Many an inventor has learned this the hard way. Tesla had his problems with the patent examiners, but today's free-energy inventor has it much tougher. The US Patent Office has even devised a Sensitive Application Warning System. (See "The SAWS Memo" on Teslapress.) SAWS flags for "special handling" any application that smacks of cold fusion or any other free-energy method.

Tesla's free-energy receiver was patented in 1901 as An Apparatus for the Utilization of Radiant Energy. The designation in the free-energy parlance of today would be space-energy receiver. (The solar panel is actually a very primitive space-energy receiver.) Tesla's patent refers to "the sun, as well as other sources of radiant energy, like cosmic rays." That the device works at night is explained in terms of the night-time availability of cosmic rays. Tesla also refers to the ground as "a vast reservoir of negative electricity."

Tesla was fascinated by radiant energy and its free-energy possibilities. He called the Crooke's radiometer (a device which has vanes that spin in a vacuum when exposed to radiant energy) "a beautiful invention." He believed that it would become possible to harness energy directly by "connecting to the very wheelwork of nature" His free-energy receiver is as close as he ever came to such a device in his patented work. But on his 76th birthday at the ritual press conference, Tesla (who was without the financial wherewithal to patent, but went on inventing in his head) announced a "cosmic-ray motor." When asked if it was more powerful than the Crooke's radiometer, he answered, "thousands of times more powerful."

how it works

Tesla's space-energy receiver works from the electric potential that exists between the elevated plate and the ground. Energy builds in the capacitor, and, after "a suitable time interval," the accumulated energy will "manifest itself in a powerful discharge" which can do work. The capacitor, says Tesla, should be "of considerable electrostatic capacity," and its dielectric made of "the best quality mica, for it has to withstand potentials that could rupture a weaker dielectric."

In my own experiments, the capacitor that performed best was an exotic pulse capacitor, rated 62 mfd, 2 KV, and an incredible

600 amperes. I ganged six of these in series-parallel for a total of 6 KV. I also tried a big electrolytic rated 10,000 mfd, 450 volts, but the pulse capacitors performed better. Needed is a capacitor rated 6 KV or better with the highest capacity available. The entire system – collector panel, conductors, capacitor, etc. – should be insulated to 6 KV or better.

Tesla gives various options for the switching device. One is a rotary switch that resembles a Tesla circuit controller. Another is an electrostatic device consisting of two very light, membranous conductors suspended in a vacuum. These sense the energy build-up in the capacitor, one going positive, the other negative, and, at a certain charge level, are attracted, touch, and thus fire the capacitor. Tesla also mentions another switching device consisting of a minute air gap or weak dielectric film which breaks down suddenly when a certain potential is reached.

Plauson's converter

Plauson's converter

Tesla's invention may have helped to inspire the many other inventors who have worked in the field of free energy. At least a dozen are on record. Let's look at one in particular. In 1921 Hermann Plauson, a German experimenter, succeeded in obtaining patents, including one in the U. S., for Conversion of Atmospheric Electric Energy.

In school, every introduction to electricity touches upon the phenomenon of so-called "static" (or electrostatic) electricity, and this is what Plauson means by "atmospheric electric energy." Static electricity is built-up charge, electricity in a raw state, and it comes easily in Nature, as evidenced by lightning and the Aurora Borealis. If you have ever seen a frictional static machine in operation, it's not difficult to imagine the tremendous potential in artificially produced static. A rotating disk type of static machine or the silk belt type, as in the Van de Graff generator,

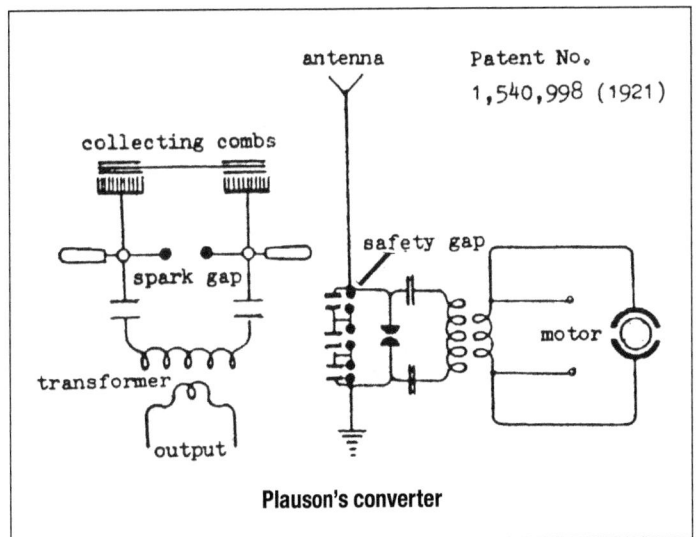

caps: left 10,000 mfd, 450 volts
right: six 62 mfd. 2KV, 600 amperes

produces discharges like those from a Tesla coil. Unfortunately, in school, the subject of static electricity is briefly touched upon and then abruptly dropped, never to be mentioned again. Electrical power sources thereafter are limited to the battery or the wall socket.

how it works

In the Plauson drawing the free energy converter on the left interfaces with a disk machine via special pick up "combs." When the static collecting disk is rotated, the combs pick up the charge, one comb going positive, the other negative. The combs, in turn, charge up their respective capacitors until a sufficiently high potential builds to jump the spark gap. The oscillatory discharge is induced into the transformer primary.

This is high-voltage, high-frequency electric energy. The familiar spark-gap oscillator has turned charge into dynamic energy. The transformer steps down the vibrating high voltage to practical levels to power lighting, heating, and special high-frequency motors.

The Plauson patent drawing to the right shows a device that works on the same principle but collects energy by means of an antenna terminal, as does Tesla's receiver. Since the higher the antenna the better, and the more area the better, Plauson favors big metallic helium balloons. Plauson says the safety gap, which has three times the resistance of the working gap, is absolutely necessary for collecting large quantities of charge. The capacitors across the gaps in the series safety gap allow for uniform sparking.

Plauson's device suggests that Tesla's might be explained in terms of electrostatics. Tesla, at the press conference honoring his 77th birthday in 1933 declared that "electric power was everywhere present in unlimited quantities and could drive the world's machinery without the need of coal, oil, gas, or any other fuels." A reporter asked if the sudden introduction of his principle wouldn't "upset the present economic system." Tesla replied, "It is badly upset already."

Appendix A
Tesla Electrotherapy

spiral ray wand

You might consider Tesla's electrotherapy to be among his lost inventions. Tesla had a hunch that, since his high-potential, high-frequency currents could be passed into the body harmlessly, "these currents might lend themselves to electrotherapeutic uses." Tesla experimented upon himself. When he was struck down in the streets by a New York taxi, he didn't deliver himself over to the medicals but dragged himself up to his hotel room where, in seclusion and with the help of his own electrotherapy, he recovered from his fractures and contusions.

Tesla never patented in electrotherapy, but in 1891 he began publishing his observations in technical journals, and seven years later we find Tesla giving a speech to the American Electro-Therapeutic Association in which he details with drawings the high-frequency apparatus he invented for this purpose, which included a Tesla coil.

Lakhovsky

Tesla's suggestions were taken up in earnest by George Lakhovsky, who perceived that the twisted-filament, coil-like structures within all living cells constitute ultra-microscopic circuits "capable of oscillating electrically over a wide scale of very short wavelengths." Lakhovsky's apparatus evolved from Tesla's. "These circuits," Lakhovsky wrote, "are stimulated by damped high-frequency currents from a spark gap. Thus each circuit of the transmitter vibrates not only on its natural frequency, but also on numerous harmonics." Here we must sing praises to the old spark gap because Lakhovsky observes that the frequency of his oscillator ranged from 750 kilocycles all the way up to 3 gigacycles! And he adds that "each circuit also emits many harmonics, which, with their basic waves, their interferences and their effluvia can reach the scale of infra-red and even that of visible light."

Lakhovsky employed spark-gap oscillators, Tesla coils, and even vacuum-tube oscillators, and he put some of these devices into patent. The Lakhovsky multiple-wave oscillator (MWO) terminates in a distinctive frequency-independent antenna consisting of a number of concentric open rings of different diameters. The MWO antenna provides full-body stimulation to the patient, who is situated a few feet distant from an antenna or between a matching pair.

Tesla's electrotherapy idea was taken up as well by Arsene D'Arsonval and Paul Ouden. One finds in the Tesla coil literature many mentions of an "Ouden coil," when a Tesla coil is obviously meant. This has perplexed some researchers who conclude Ouden's coil had to be special, but he had just made it a safer apparatus. (The Ouden circuit grounded the bottom of both primary and secondary, but this feature can be found in Tesla's circuits as well.) Perhaps, as Tesla's name became taboo in the media, writers and editors chose to call the device by Dr. Ouden's name to play it safe.

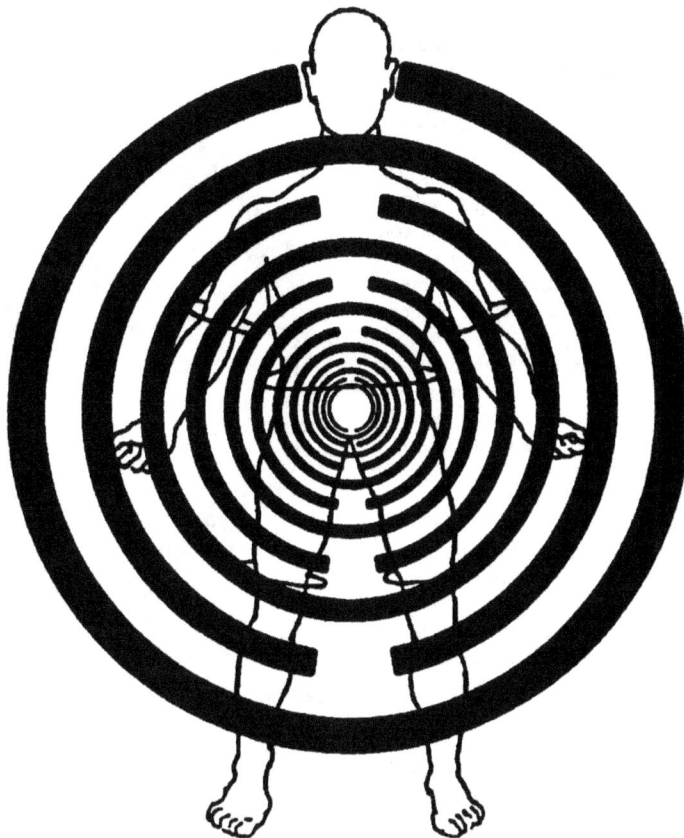

from MWO Handbook

medical secret

Lakhovsky called his book *The Secret of Life*, no less. The ability to electro-stimulate living tissue at the subcellular level and thus energize the life force within has huge medical implications. Organized medicine (which works hand-in-hand with pharmaceutical corporations, which in turn work hand-in-hand with the mass media) distracts the public from the observation that the myriad diseases that afflict us could stem from a fundamental condition, the weakening of the "life-force" or of the "heart of health" (to use some old terms rather then the "immune system" of contemporary AIDSpeak).

In modern medicine each and every disease, disorder, and (more recently) "syndrome" is assigned its own particular pathological designation, its own symptomatology, its own etiology (cause), and, if possible, its own particular medical specialty. Thus particularized, each disease can have its own therapy, be it a vaccine, an antibiotic, an anodyne, a surgery, or whatever, and may even have its own medical specialist. One of the most hugely profitable industries on the planet has developed out of this distraction and brainwash that passes for modern healing. The scam has gone so far now that researchers invent diseases and syndromes by definitional contrivance, even when no distinct and separate symptomology or etiology exists. (Example: so-called AIDS)

Lakhovsky proposed that exposure to a blend of higher frequencies stimulate the cell's life force, restoring vigor and balance. The vibrational responsiveness of living cells suggests a whole new medical panorama in which electric waves, both natural and man-made, exercise influences both healthful and malignant upon the body's cellular oscillatory balance.

According to Lakhovsky, treatment with the multiple-wave oscillator mobilizes the body's own self-healing reserves. Thus the range of diseases that can be treated is infinite.

Degenerative conditions develop when the body's self-healing reserves lose their power. Infections, cancers, inflammations, skeletal degeneration and organ dysfunctions then develop, but often such conditions can be reversed if these reserves are revived. Even fractures and cuts can be healed in a fraction of the normal time. Neural dysfunctions, from headaches to deafness to paralysis, can be normalized. The MWO has been used successfully to treat arthritis. Can any such cure-all really exist? If there is a generalized life-force enhancer, then the answer is yes, and this may be it.

violet-ray

An allied mode of MWO-style electrotherapy is the violet ray. The violet ray wand is another convenient means of translating electric energy into the body, but in a more focused, localized mode. A low-pressure inert noble gas, such as argon, contained in a glass bulb or tube, is electrified by high-potential, high-frequency Tesla currents generated by a spark-type Tesla coil. The wand emits, when brought into contact with the body, an electric ray, seen as a orange, reddish or violet beam, a fascinating phenomenon to watch.

Tesla himself used such a revivifying ray daily. While the MWO was never mass-produced, the violet-ray machine was actually commercially manufactured, and it became a fixture in many a doctor's office and in many homes. Made available to the general public by a number of manufacturers in the 1920s and '30s, one could mail-order the device from a Sears catalog. Not surprisingly, the advertising made sweeping cure-all claims. Eventually medicine organized to suppress this threatening alternative to its official line, which it labeled "quack." But for a time both MWO and violet-ray flourished, and to such a degree that it still cannot be completely stamped out.

Like Tesla technology generally, this high-frequency electric healing technology still persists today world-wide and underground. Borderland Sciences made a big contribution to the perpetuation of the technology with the publication of Tom Brown's *MWO Handbook* in 1986, which is now in its fourth edition. The violet-ray machines employed compact spark-gap oscillators or spark-driven Tesla coils to generate the currents.

The most common ray-tube electrode was in the form of a wand with a flared end, but ray-tubes were also available in a wide variety of blown-glass shapes designed to accommodate any contour of the anatomy and to fit into any bodily orifice.

violet-ray-today

When appropriate inexpensive violet-ray electrodes became difficult to come by, experimenters found an alternative in an off-beat argon UV night-light bulb called the AR-1. I used one of these for many years. Fortunately, one does not have to resort to such improvisations, for today a few dealers continue to sell a mushroom violet-ray wand for as low as $7.00. Violet-rays have become a fashion in sex-fetish circles, which enhances the availability and affordability of the technology, as does the expanding popular interest in Tesla-coil building. However there are some violet-ray and MWO appliances which are outrageously priced waiting to rip you off on the web.

The seer and healer Edgar Cayce recommended violet-ray electrotherapy for his patients in some 900 readings and for a tremendous variety of conditions, including arthritis, baldness, circulation problems, nerve, spinal and debilitation problems, sprains, eye disorders, and even demonic possession.

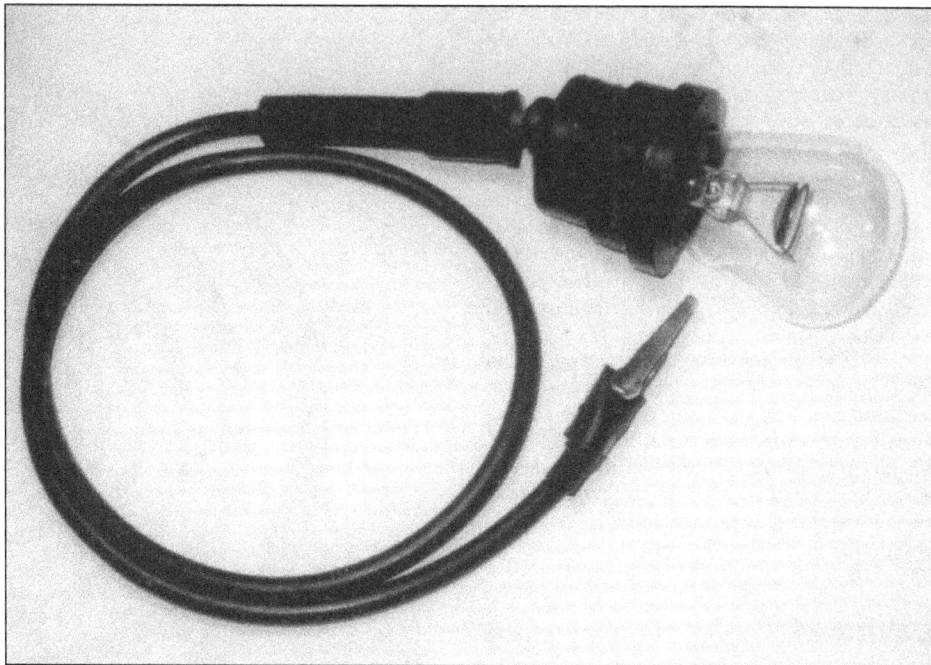

AR-1 bulb

Tesla-coil

I have yet to build the concentric-ring antenna or the MWO, but I have plenty of direct experience with the old AR1 violet-ray bulb, which I recently replaced with a wand. The violet ray has been my way of translating Tesla currents into the body. I connect the bulb or wand to the terminal of a spark Tesla coil. The Tesla coil can be a small low-power unit or a larger one moderated to low power by a variable-voltage transformer (variac).

Immersion of the Tesla-coil in oil enhances the effect. I have experienced the value of oil immersion in the "recipe" oil coil described in detail in Part 3, "Tesla DIY." For electrotherapy I've found it superior to any open helical secondary. It's power can be felt, for it generates in the tissue a greater heating (diathermic) effect.

Must the Tesla coil be spark-gap type rather than electronic? As an experiment, I drove my violet ray with the solid-state pulse unit (also detailed in Part 3). Unlike spark, which generates a broad spectrum of frequencies, the electronic driver supplies a discreet frequency of electric vibration to the wand. This produces a steady intense ray of different coloration, no crackle, and very hot. The diathermic effect is powerful, nearly burning, and lingers in the tissue for some minutes after use. Interesting experiment; this strange ray may have a use. But, for electrotherapy as I know it, I'll stick with the tradition of spark.

I've also built and used extensively a little portable MWO coil designed by Bob Beck (12-volt solid-state-driven ignition coil, spark gap of auto points, tiny one-inch diameter secondary).

Lately I prefer larger, hotter stationary coils, but powered down with a variac. My latest spark gap is also made from auto ignition points. The transformer (neon) is attenuated by the variac and is rated only 5 kv, 20 MA. The spark gap is about .004". The application time for ray bulb or wand can be from one to thirty minutes. Violet-ray devices should always be adjustable for intensity, the output being reduced when the diathermic effect is felt to be too hot.

Tesla coils are inimical to computers and other solid-state electronics. My violet-ray apparatus often creates a little nuisance by erasing the recorded greeting in the solid-state answering machine across the room. Tesla-coil devices, including the violet-ray, can disturb or even destroy the operation of any sensitive solid-state device, including the cardiac pacemaker, so avoid Tesla currents if you have such an implant. By the same token, if I found myself implanted with an unwanted electronic implant, such as an RFID chip, the violet ray would be applied as my first effort to destroy it. Particularly vulnerable within integrated-circuit chips are tiny capacitors whose dielectrics can be punctured by excessive voltages. The Tesla coil powers the violet-ray with voltages in the many thousands, safe for you at the high frequencies employed, but possibly fatal to any implanted mini-electronics.

ozone

Holding the wand to my face with one hand, I can grasp a fluorescent tube with the other, and the tube flickers. Electrifying. This is a way of experiencing the Tesla coil viscerally. As I put the fluorescent tube in the circuit, the increase in capacitive terminal load pulls up the voltage, and more so if I ground one end of the fluorescent. The ray-wand crackles. The wand must be held firmly to the skin or little sparks may arc from its corona, producing an irritating tickle. Fresh ozone is in the air. The ozone produced is one of the touted benefits.

Says electrotherapy inventor H.G. O'Neill in a patent of 1899 (No. 628,352): "Ozone in this nascent form is very much more energetic than in a free state and produces instant oxidation of all diseased matter. This form of asepsis is applicable to the entire tract of a wound or diseased surface at any depth. It is fatal to germ life and affords a means of internal asepsis." Others have touted the release of heat in the tissues (diathermy), as well as an increase in the local blood supply and an increase in the metabolic rate.

Tesla's other electric-ray explorations

Tesla was in the vanguard of X-ray development. He corresponded with Roentgen. The x-ray is a high-voltage vacuum tube driven by an induction coil or a Tesla coil.

Tesla invented an open-ended vacuum tube known in the lore as the "Tesla death ray." To maintain a vacuum in a glass tube having an outlet to open air is a feat that Tesla accomplished by the understanding of fluidic engineering, which we have seen in his valvular conduit.

Tesla's fascination with the ray devices called the radiometer and the rotating brush are discussed elsewhere in this book.

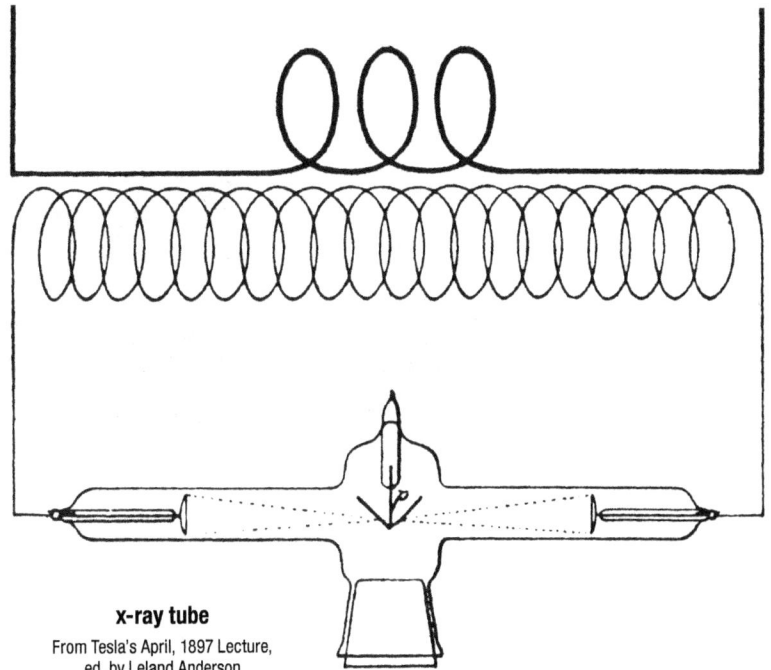

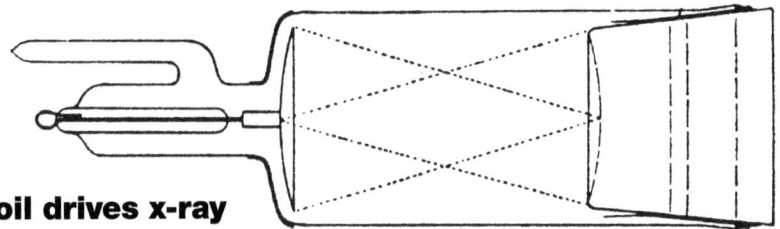

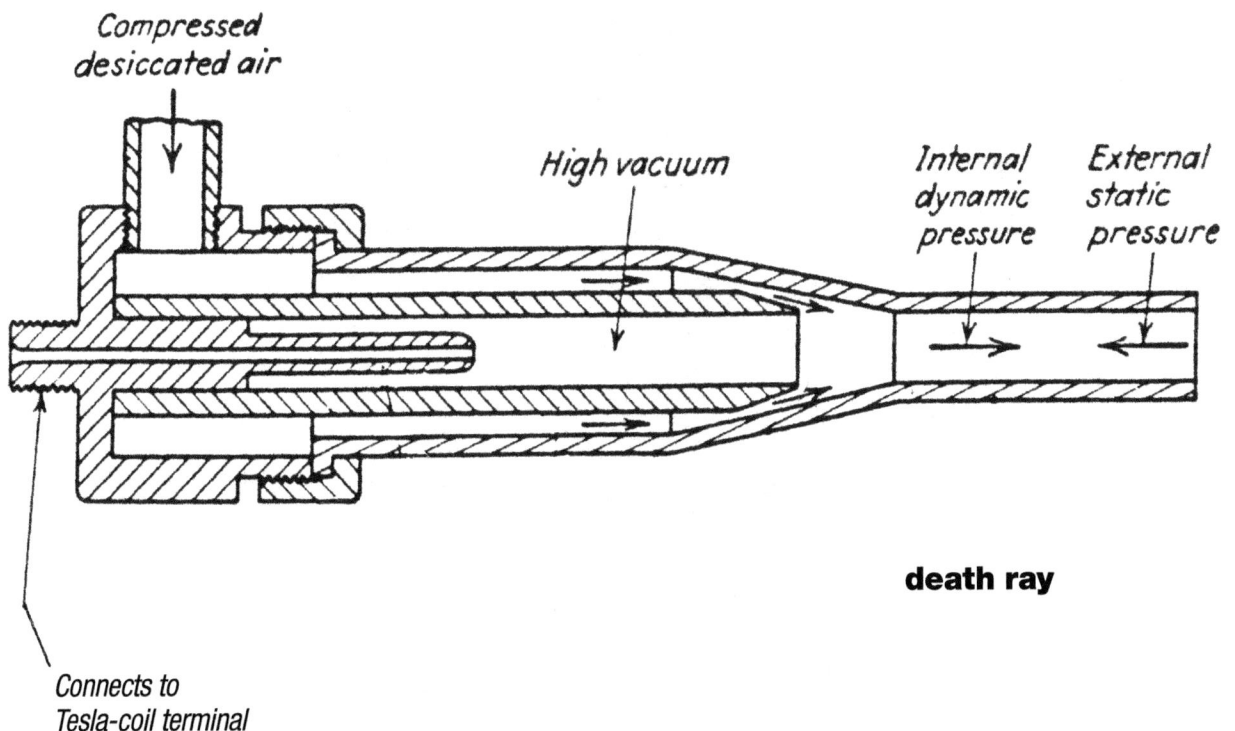

x-ray tube
From Tesla's April, 1897 Lecture, ed. by Leland Anderson

Tesla coil drives x-ray

Compressed desiccated air

High vacuum

Internal dynamic pressure

External static pressure

Connects to Tesla-coil terminal

death ray

Part 2
Radio Tesla

Introduction

When radio was born around the turn of the 20th century, various inventors, who are not celebrated today, created their own peculiar radio technologies, which are largely ignored today. Among these inventors is Nikola Tesla, although there are others, like Nathan Stubblefield and Mahlon Loomis, who are even more obscure. The radio technology that is peculiar to Tesla, though it got a few years of public exposure in its time, gets even less acceptance in today's technology than Tesla's disk turbine, his Tesla coil, or his high-frequency lighting, and Tesla's radio is as taboo in official science as his wireless power, which works on the very same principles.

In 1943, only a few months after Tesla's death, the U.S. Supreme Court, yielding finally to the pressure of a suit fought over many years, declared that Tesla's radio patents were among those that had been infringed upon by Marconi and thus, in effect, wrote into the official record Tesla's status as a founder of radio. This was a purely symbolic victory, for Tesla's radio was suppressed, and the radio technology that developed is distinctly different in many essential respects.

During the period of radio's most rapid growth (1915-1940), Tesla watched quietly from the sidelines, for by this time he had fallen out of favor with the media, or rather with the establishment that controls it. Still, some of his comments have made it into the published record. In 1927 Tesla said that broadcasting "is now carried out with unfit apparatus and on a commercially defective plan." Of radio technology generally, Tesla said in 1932 that "the transmitting and receiving apparatus is ill-conceived and not well adapted for selection. The transmitter generates several systems of waves, all of which, except one, are useless. As a consequence only an infinitesimal amount of energy reaches the receiver, and dependence is placed on extreme amplification?

"Radio experimenters of this age," Tesla said of the hams of 1934, "are following ancient theories." By this he meant backward theories. Tesla's favorite backward theorist was Heinrich Hertz, who saw the phenomenon of radio as some kind of straight-line radiation akin to light. Tesla said Hertzian theory, which is still in vogue today, was "one of the most remarkable and inexplicable aberrations of the scientific mind which has ever been recorded in history."

This was not a reckless statement, for Tesla reports that he had carefully reviewed Hertz' experiments, had conducted comparative tests with his own brand of radio, and had come to a different set of conclusions. "1 considered this so important," said Tesla, "that in 1892 I went to Bonn, Germany to confer with Dr. Hertz in regard to my observations. He seemed disappointed to such a degree that I regretted my trip and parted from him sorrowfully." Tesla made subsequent tests in 1900 with the same results and kept abreast of articles on Hertzian radio-telegraphy, which, he said, always impressed him "like works of fiction." Official science generally consists of works of fiction, I say.

In this book I attempt to break down Tesla's radio into a set of specific principles and to survey the whole of radio technology from Tesla's perspective.

For the core material on Tesla's radio, what you have is one fat hardcover book, *Colorado Springs Notes*, eight U.S. patents, and a few magazine articles by Tesla himself, one of which is *The True Wireless* (see Appendix B.)

The biggest single source, *Colorado Springs Notes*, embraces only one year (1899) of Tesla's many years in radio, albeit a very intense year. Although this work is rich in information, it represents only a tiny fragment of Tesla's total legacy, which has been said to amount to some 100,000 documents, including 34,552 pages of scientific material and 5,297 pages of technical drawings and plans. Though much of Tesla's radio research took place in New York City in years prior to his Colorado adventure, there is no published volume called "New York Notes."

suppressed information

Colorado Springs Notes itself has a bizarre publishing history. Tesla's papers, confiscated by the U.S. government upon his death, surfaced years later in a Tesla Museum in Belgrade, or at least a large chunk of the expropriated material did. When the Museum, thirty or so years after Tesla's death, published the Colorado fragment, it did so in Serbo-Croatian, a curious choice since Tesla, though of Serbian origin, wrote these notes in English. What the Museum published as *Colorado Springs Notes* in 1978 (under the imprint NoLit) had been translated back into English from the Serbo-Croatian. The translator, in his appended notes, tends to discredit Tesla's version of radio. The discrepancies between the original and the translation have prompted a study by pioneering Tesla scholar John Ratzlaff called *Serbo-Croatian Diary Comparisons.*

These frustrations noted, you still have to be grateful for the relatively abundant documentation of Tesla's radio relative to what you would find in print on any other deviant radio pioneer, such as Loomis or Stubblefield.

To put Tesla in perspective, I have also read extensively in the literature of conventional radio, and, by way of grounding this information, have invested perhaps an inordinate amount of time in designing, building, and experimenting with elemental transmitters and receivers, some of which are illustrated here. (Part 3: Tesla DIY shows you how to build a regenerative receiver.)

In this book I share with you many illustrations of radio transmitters and receivers of the past, because most have pertinence to Tesla and also because these antique technologies are at risk of disappearing down the memory hole. These transmitters and receivers are elemental devices that teach us that powerful radio technologies are more simple and accessible than one might imagine.

Tesla's radio principles

Tesla believed that radio was conduction, not radiation. A radio transmitter should put out high voltage impulses sharp and sudden. Tesla's radio was grounded rather than aerial. He saw radio as most potent in the lower frequencies. For both transmitting and receiving, the magnifying of effects, he believed, should be achieved with resonance and capacitive discharge rather than with many stages of amplification.

Tesla believed radio was a disturbance of a pervasive medium called the ether. Rather than a radiating antenna, he employed an aerial capacity. Tesla's radio inventions, like his others, show a striving for simplicity rather than complexity. He eschewed the miniaturization fashionable today, conducting his experiments on a fearless scale. Fundamental to resonant tuning was Tesla's quarter-wave principle.

Tesla planned, promoted, and began construction of a World System of radio founded on these principles. On these same transmitting principles, and with about the same equipment, Tesla experimented with his wireless power. This was a radio-like system that was to be a successor to the wired grid, which was built on Tesla's own 60-cycle alternating current patents.

Tesla's wireless power system (also patented) was designed to deliver, by means of earth-electrical vibration, power sufficient for industrial demands. Tesla boasted of the simplicity of his wireless-power receiving system: "Any person skilled in the mechanical and electrical arts can utilize to advantage the practical applications of the system," by which he means that any person so skilled can construct the apparatus, receive the energy, and put it to use. Thus Tesla's system short-circuits the techno-priesthood, inviting suppression.

Tesla is on record as saying that he lit up 10,000 watts worth of Edison bulbs wirelessly 26 miles from his magnifying transmitter at Colorado Springs, but if Tesla wrote an entry on this remarkable experiment, it is not to be found in the English version of the published *Notes*. Whatever the viability of Tesla's wireless power, his conviction that such a thing was possible testifies to his belief in the special power of his peculiar radio technology.

CHAPTER 1

High Voltage, Sudden Pulse

Tesla-style radio transmitting applies high electromotive force to the aerial-ground system, "hundreds of thousands or millions of volts," said Tesla. These extraordinary-sounding pressures can be most easily achieved using the Tesla coil, and fact is that Tesla's transmitters were Tesla coils.

The Tesla coil, with the addition of a resonant third coil and a few other alterations, becomes Tesla's magnifying transmitter, which Tesla boasted "would enable the obtainment of practically any emf, the limits being so far remote that l would not hesitate to produce sparks thousands of feet long in this manner."

In modern vacuum-tube and transistor transmitters, potentials in the hundreds of thousands, much less millions, of volts are unheard-of. Rarely does the voltage in vacuum-tube-transmitter plate circuits exceed ten thousand, and it is usually under 1500. Antenna voltage rarely exceeds 1000. In calculating the performance of a modern transmitter, focus is on antenna current, not voltage.

sudden pulse

In addition to the high voltage, Tesla-style transmission requires sudden pulses, in Tesla's words, "an immense rate of momentary energy delivery." This also comes easily to the Tesla-coil transmitter, driven as it is by the sudden capacitive discharges of Tesla's spark-gap oscillator.

Tesla's transmitters

"With electromotive impulses not greatly exceeding 15 to 20 million volts, the energy of many thousands of horsepower may be transmitted over vast distances, measured by many hundreds and even thousands of miles." So writes Tesla in a basic radio and wireless-power patent. Among the uses for this electromotive energy Tesla cites the transmission of "intelligible messages to great distances," power for "industrial purposes," the production of nitric acid and fertilizer by electrifying nitrogen in the atmosphere, and the illumination of "upper strata of air," this last item referring to Tesla's outrageous scheme to light up areas of the planet at night by beaming oscillating electric energy into the sky.

"It is a form of oscillator of great simplicity," Tesla wrote. "A coil of high self-induction is connected in series with a condenser, and across the condenser is placed a break generally in series with the primary of the coil. Very sudden discharges are produced when using a stream of electrolyte or mercury to effect

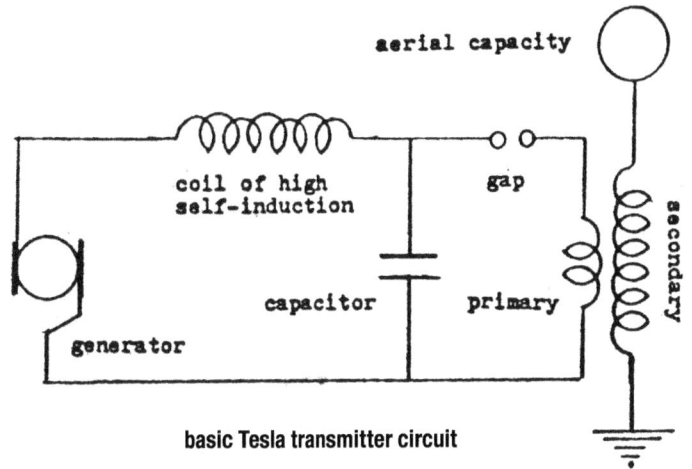

basic Tesla transmitter circuit

short circuit," says Tesla, who patented circuit controllers using conductive fluids such as mercury.

engineering

The engineering of the Tesla transmitter is the engineering of the Tesla coil. The following guidelines are from Tesla's patents and Colorado notes:

The primary coil's inductance, and its resistance, should be as low as possible. Just one turn of heavy double cable sufficed for the 51-foot diameter primary of Tesla's Colorado transmitter.

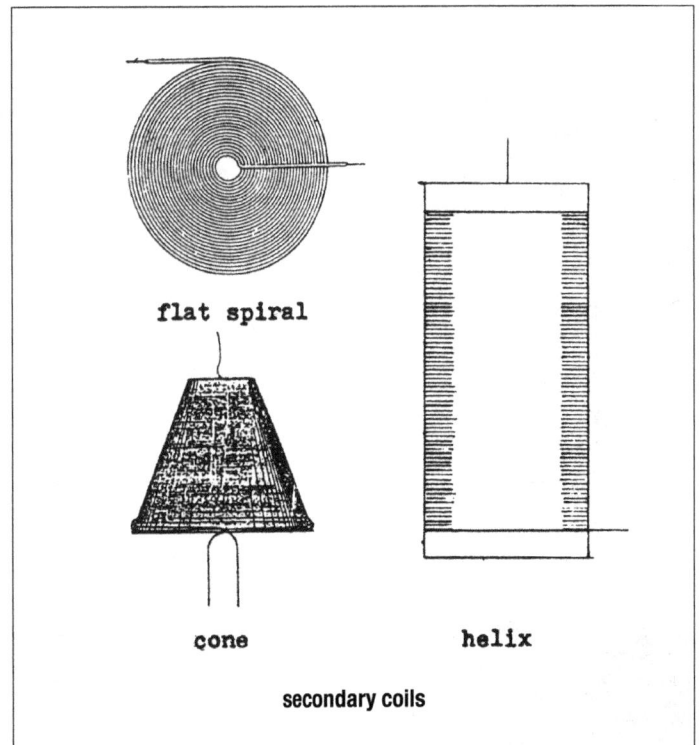

secondary coils

The primary and secondary should have equal weights of metal. This principle helps determine the thickness of the primary conductor.

Of the secondary, Tesla wrote that its resistance, too, should be as low as possible, but the self-induction should be as high as possible. The flat spiral secondary of the coil mentioned above had 50 turns of number 10 wire. The helical secondary of the Colorado transmitter had 24 turns of number 8 on its 51-foot diameter.

A loose coupling between primary and secondary is desirable. "'The mutual inductance should be small so as to permit free oscillation," said Tesla. A coil oscillates most vigorously when it is "inductively free of other circuits." The "extra coil" or tertiary of the magnifying transmitter can achieve great amplitudes because it is free from the damping effects of other coils. The extra coil can be located some distance from the coil that stimulates it. Within the Colorado transmitter's 51-foot diameter primary-secondary coil system, Tesla would set down a tertiary, not necessarily centered. One Colorado photo shows several of these extra coils, ranging in diameters from seven inches to eight feet, placed here and there within the 51-foot enclosure, each spouting sparks.

Loose coupling for a more free and swinging oscillation is an engineering objective behind the flat-spiral secondary design. As the coil spirals inward, the turns become more and more remote from the primary and its inertial influence. Also, since the secondary takes on higher voltage as it spirals inward, the outer turns nearer the primary are at relatively low potential, and this helps to thwart arcing between the coils. A cone-shaped secondary offers similar advantages and is also considered superior to the customary helical.

The magnifying effect of the interacting coils, Tesla said, "is directly proportionate to the inductance and frequency and inversely to the resistance of the secondary system." This need for low resistance invites the builder to consider conductors superior to conventional copper wire. A video from Borderland Sciences demonstrates a Tesla-style transmitter with a flat-spiral secondary made of silver-teflon coaxial cable; the shielding is the conductor. The same coil has a primary of bronze strap.

Tesla tuned his transmitters for "maximum rise of emf on the excited system," as indicated by a small incandescent bulb connected in a single loop around the final coil. Tuning was aided by a regulating coil and by an adjustable spark gap in the ground circuit or by a gap shunting part of the secondary. He observed that "the tuning is remarkably exact, 1/8 turn of the self-induction box reducing the effect very much." He said, "It becomes easy to locate the maximum rise within one-quarter of one percent." He described his transmitter's tuning as "very sharp."

spark

Spark radio is high-voltage, sudden-pulse. Spark is as old as radio itself. Radio was supposedly discovered by somebody wondering about spark noises in a telephone. Hertz' experiments used spark. The induction coil with spark gap is the first radio transmitter. The next advance put a capacitor in the circuit, as in Tesla's spark-gap oscillator. This Tesla invented not for radio but for an early high-frequency lighting system. For radio, Tesla used his all-purpose oscillator, the Tesla coil.

As radio moved into the future, leaving Tesla behind, the spark transmitter that came into general use was not a Tesla coil but a truncated version in which the magical resonant secondary was lopped off. The power supply was an induction coil which one could build or buy. One manufacturer advertised eleven models providing sparks from 1/4 inch to 8 inches. Antenna and ground were connected to a primary-like helical coil called a transmitting helix.

The rarity of true Tesla-coil transmitters in the literature suggests that this potent oscillator may have been a well-kept secret even in the spark era.

spark forever?

"Using spark allows obtaining of great suddenness," said Tesla. No modern tube or semiconductor can match disruptive discharge for rapid switching of high energy without blowing apart. The history of radio reports that operators were satisfied with spark. "The pervading spirit was one of complete complacency with regard to the technical status of the art," according to one radio historian.

A 250-watt spark set went 300-400 miles, and with spark that power was easy to come by. The circuitry is simple, the parts easy to find even today for those in the know about Tesla-coil building. If you needed to throw together a telegraphic radio signaling device in a pinch, spark might be the way to go, and even voice modulation is possible, as we shall see.

Radio operators, both commercial and amateur, resisted mightily the movement to suppress spark and to replace it with tube oscillators, which they saw as costly, complex, and relatively ineffectual. The anti-spark movement was imposed from above by industry and government. Imprinted on some of the QSL

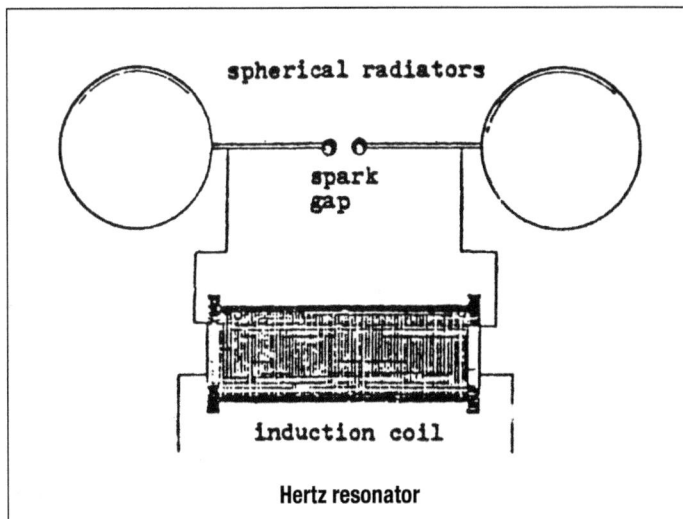

Hertz resonator

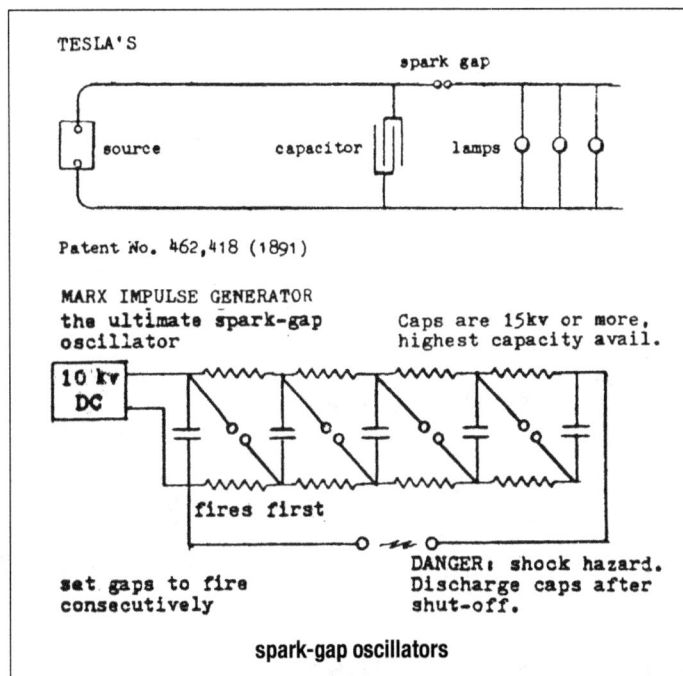

spark-gap oscillators

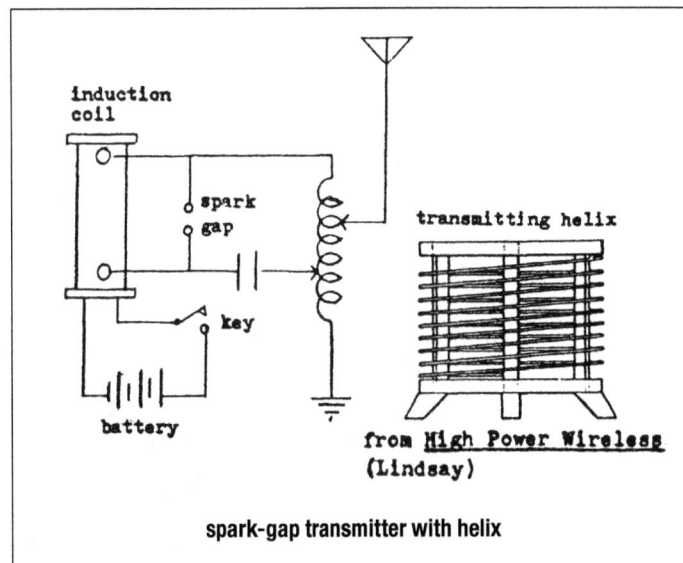

spark-gap transmitter with helix

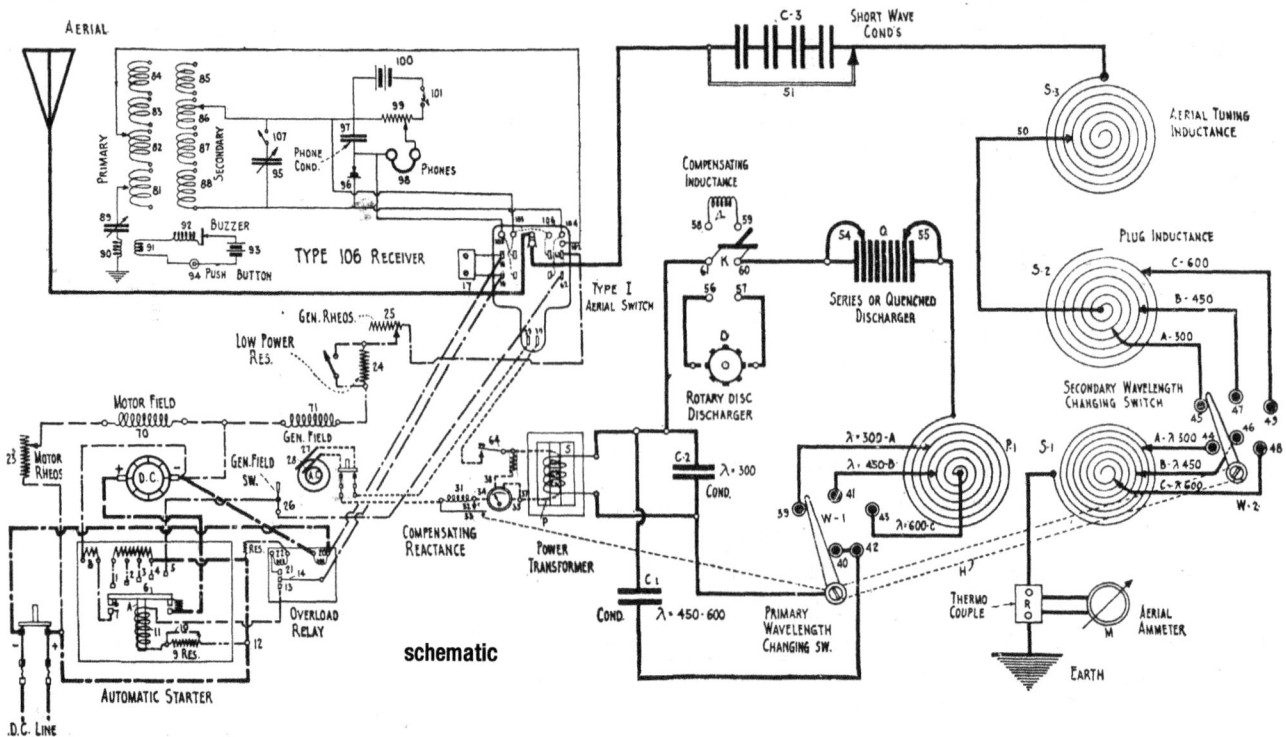

schematic

Spark Transmitter

**Marconi marine
2 kw, 500 cycles
circa 1917**

*from Practical Wireless Telegraphy
Elmer Bucher, 1917*

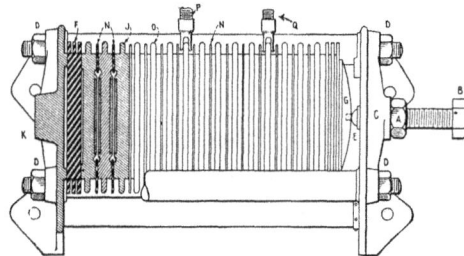

Marconi quenched gap

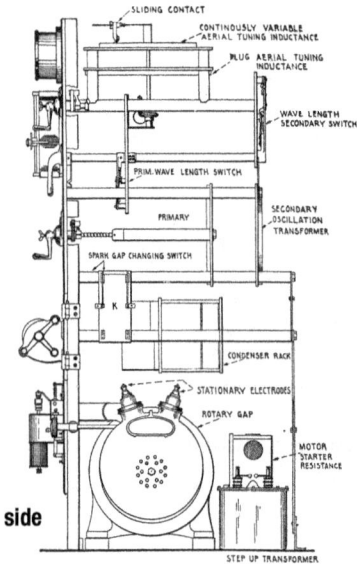

side

rear

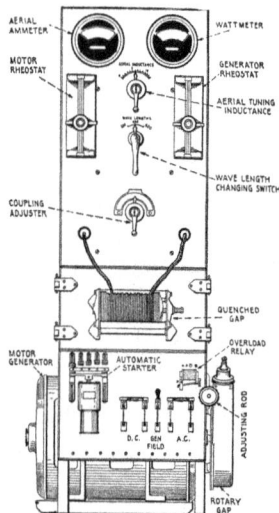

front

cards hams sent through the post to confirm contact was the slogan "spark forever."

Resistance continued into the 1930's, and, who knows, there may still be a few covert spark operators out there in the night.

etheric ghetto-blaster

The untuned emergency transmitter shown below is a crash-signaling device. In my mischievous youth, I built a portable crash spark-transmitter and used it to torment the next-door neighbor. A six-volt lantern battery drove a Model-T spark-coil (much like the induction coil shown) which fired into a simple spark gap connected to a coat-hanger antenna. An untuned spark transmitter emits a crash disturbance over a broad range of frequencies, including those of old broadcast-TV channels 2-13. In my experiments upon neighbor's TV, the disturbance from my little crash-transmitter contorted a coherent TV image into a mess of slant lines. When neighbor rose to adjust his deranged picture, I would switch the device off. When he sat down and became comfortable again, the crash signal would resume. Such a universal jamming device might be useful in certain strategic situations. One could improve upon my primitive etheric ghetto-blaster by substituting for the coat-hanger a frequency-independent antenna, like a yagi or the MWO antenna shown in Appendix A above. For the Model-T spark-coil substitute an automotive ignition-coil pulsed by one of the drivers shown in Part 3.

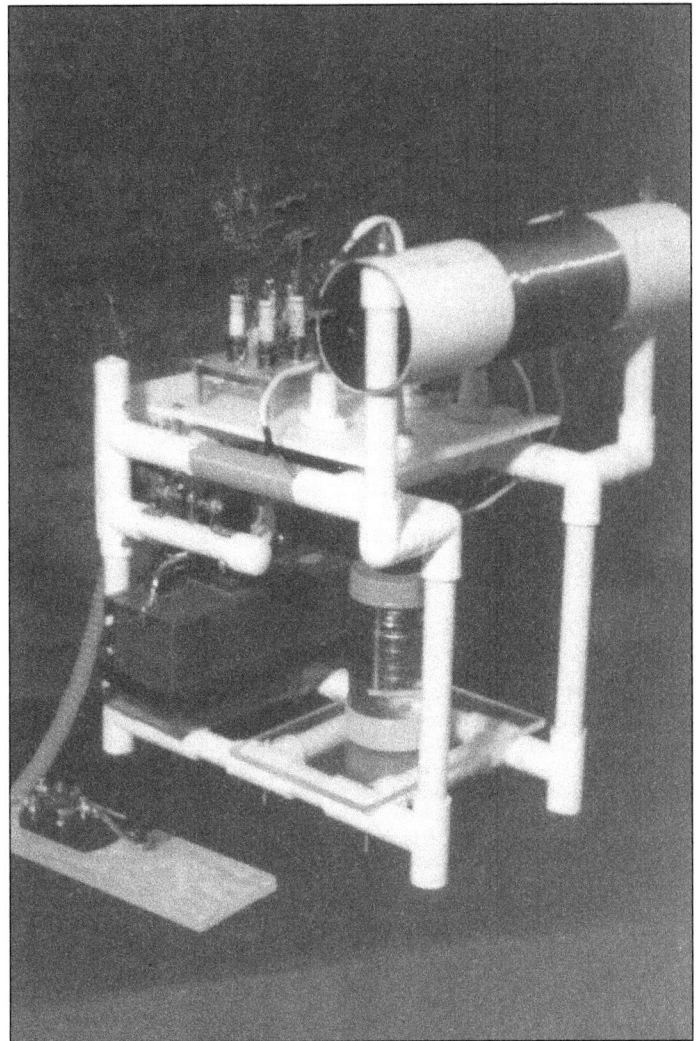

experimental spark transmitter

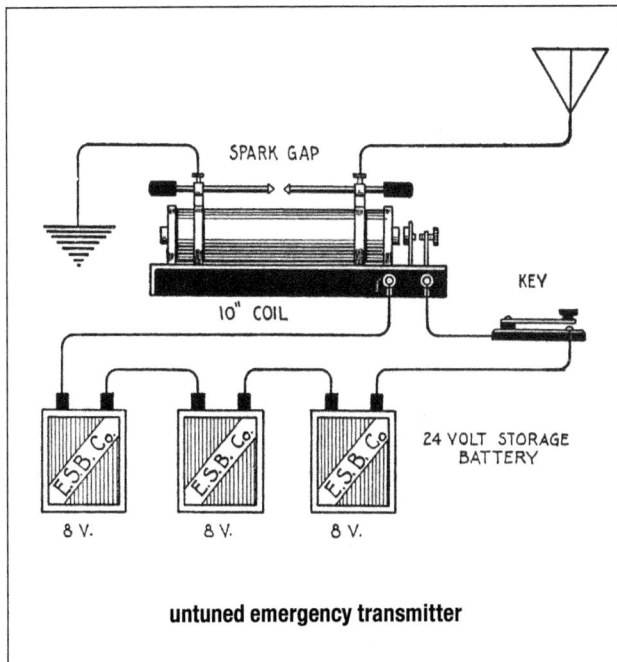

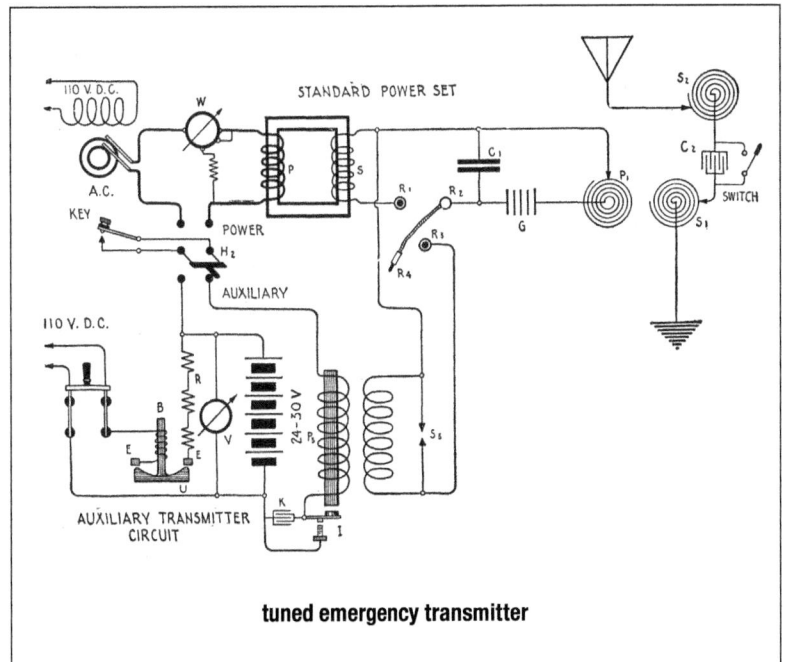

untuned emergency transmitter

tuned emergency transmitter

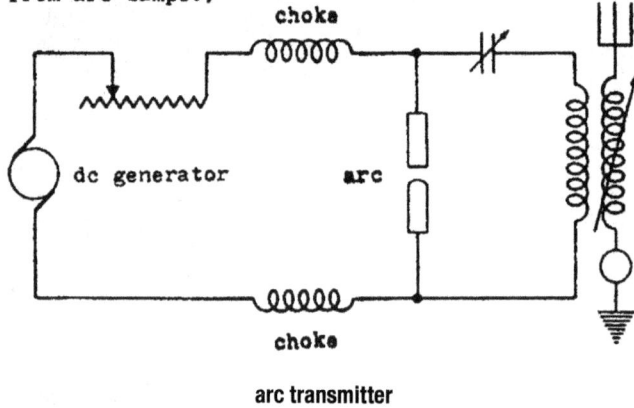

Any electric arc is an oscillator. (Early radio experimenters listened to signals from arc lamps.)

arc transmitter

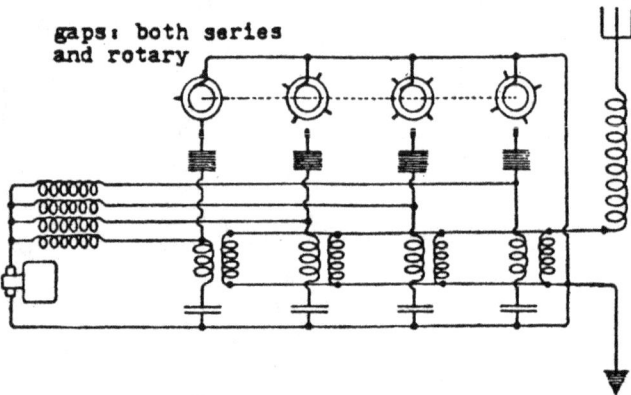

gaps: both series and rotary

continuous-wave spark transmitter

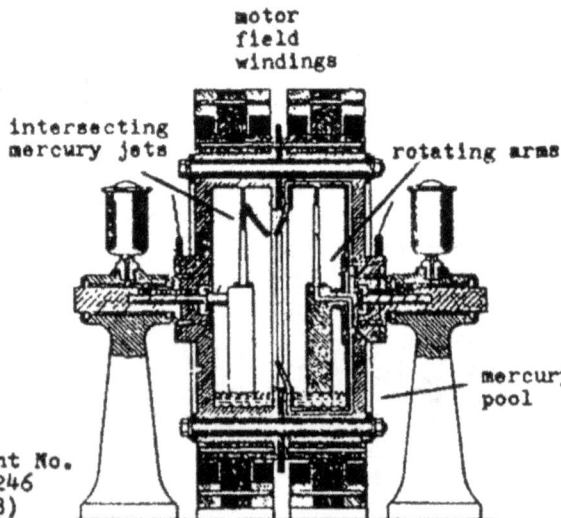

motor field windings

intersecting mercury jets

rotating arms

mercury pool

Patent No. 609,246 (1898)

Tesla's mercury break

cleaning up spark

The problem imputed to spark was broadness of signal and harmonic noise in radio bands which were becoming progressively more crowded with competing communicators. Desired was a sharp, clean signal that occupied only the thinnest slice of the busy radio spectrum. There is nothing in the FCC Code for amateurs that prohibits spark itself. The Code does set strict emission standards that the rudimentary transmitter of the spark era had trouble conforming to.

The common historical impression is that spark transmitters were uniformly and inevitably crude and that the new vacuum-tube transmitters, using feedback oscillation, came along like white knights to clean up the spectrum. Fact is, though, that toward the end of the spark era transmitters using evolved spark technologies were developed, manufactured in numbers, and saw long service on land and sea, sharing the bands with the incoming tube and arc transmitters. Perhaps harmonics were a problem, but these advanced spark rigs produced relatively clean and narrow fundamental signals, and they produced the smooth, damped, "continuous waves" boasted by the promoters of the new tube transmitters. Even after ships were routinely equipped with tube transmitters, some radio operators insisted on keeping on-board their old reliable units with spark.

The engineering of superior spark transmitting focused on improving the quenching (the shutting-off) of the gap. This is the technology of airtight series gaps, magnetic blow-outs, and rotary gaps. The perfecting of the spark gap got a lot of attention from Tesla, and perhaps his patents were studied by radio engineers, but not necessarily.

Tesla anticipated the push-pull circuit common in amplifiers and transmitters. Tesla ultimately replaced the spark gap altogether and achieved disruptive discharge with ingenious high-energy

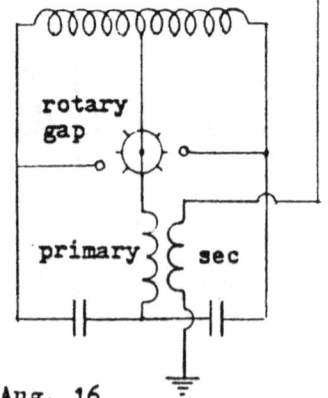

supply transformer

rotary gap

primary   sec

Colo. notes, Aug. 16

Tesla's push-pull oscillator

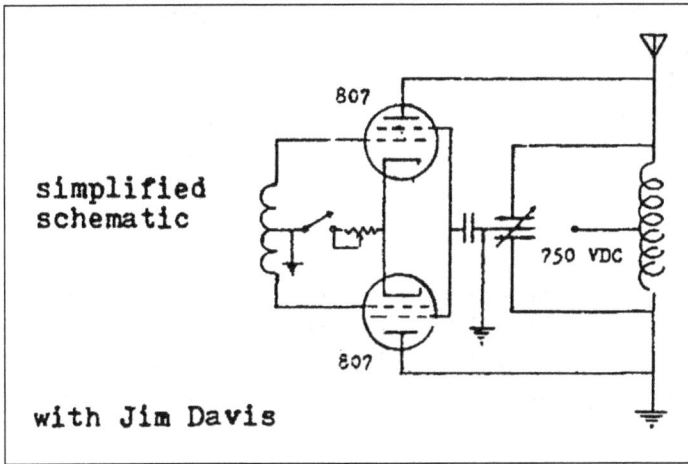

simplified
schematic

with Jim Davis

author's push-pull transmitter

mechanical "breaks," including rotary mercury switches and mercury jets. Tesla said that his transmitters produced continuous waves. Evolved spark transmitters used synchronous rotary gaps which coordinated the charging rhythm of the capacitor with the frequency.

A feature of Tesla's magnifying transmitter design that discourages undamped waves is the single-turn primary. This reduces induction in the primary circuit so that unwanted oscillation could not occur there.

The same provision shows up in the only magnifying transmitter circuit I have seen in radio outside of Tesla's. Called the impact transmitter, this circuit from the spark era featured quenched gaps in tandem. The name belies an understanding of Tesla-style sudden-pulse transmitting. Knowledge of the impact transmitter, patented by Oliver Lodge, was never widely circulated to the public and remained a secret.

The sudden-pulse spark-gap oscillator could be used to drive a microwave cavity resonator.

Could a self-powered spark transmitter be created from a static machine?

Can you modulate spark?

The general impression is that spark is strictly telegraphic, but the fact is that spark transmitters can be modulated. For telegraphy, tone modulation is inherent in spark, the tone being determined by the frequency of the gap. Rotary and series gaps produce characteristic tones that help telegraphic signals cut through static noise. (Tube and transistor transmitters require separate audio-oscillator circuits to impose a tone, yet another complexity.)

Voice modulation required that any spark-gap sounds be eliminated, so the frequency of rotary spark dischargers had to be raised to at least 5000 sparks per second.

Voice modulation is actually possible with a simple two-element spark gap. Aircraft of WWI communicated via phone transmitters employing the Chaffee gap, an airtight gap with electrodes of copper and aluminum. The bimetallic Chaffee gap quenched so capably that it enabled modulated voice transmission with great clarity. Says a radio manual of the period: "The practically instantaneous reestablishment of the high initial

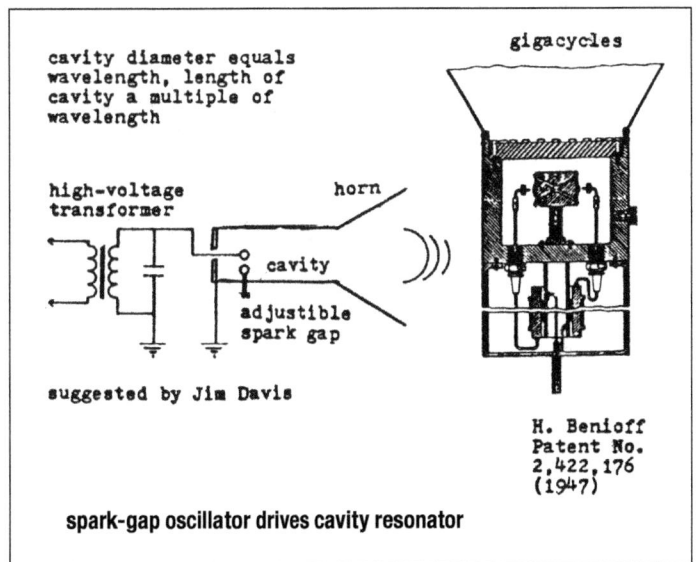

impact transmitter

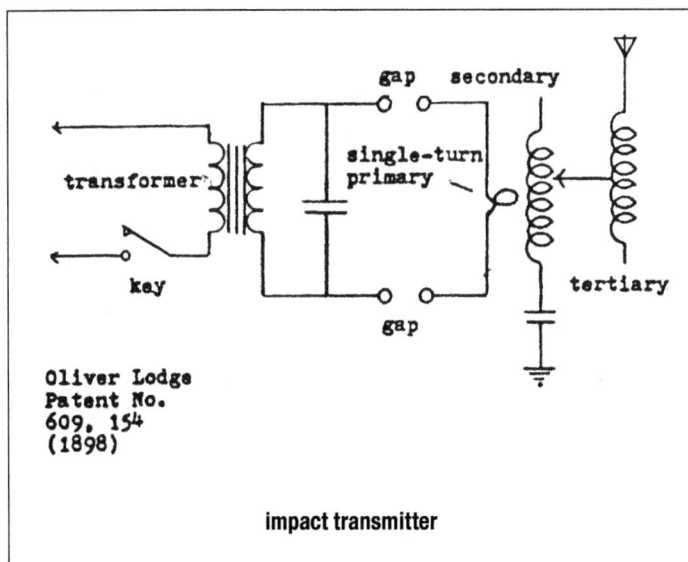

spark-gap oscillator drives cavity resonator

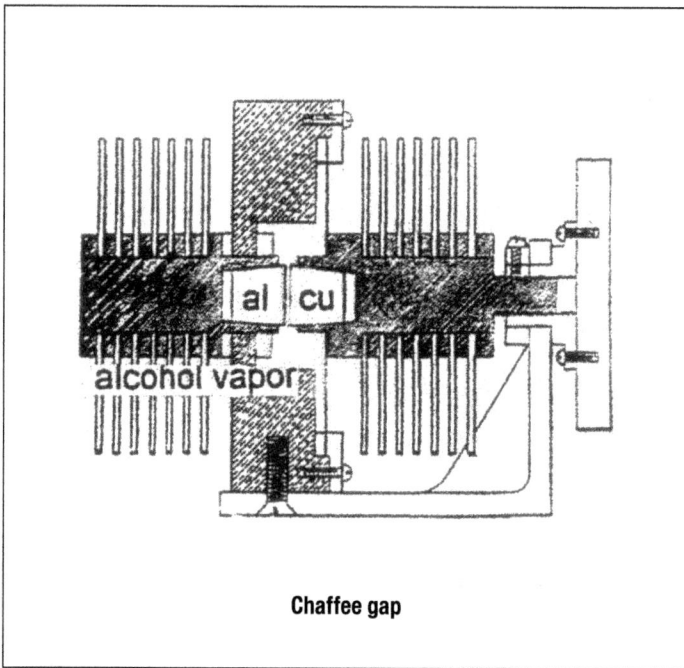

Chaffee gap

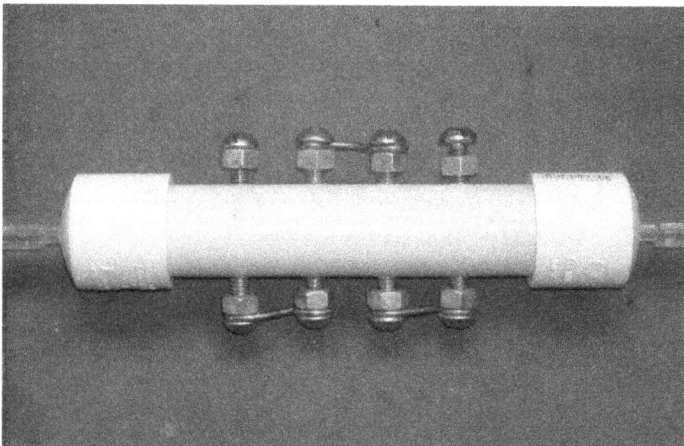

airtight series gap with gas fittings

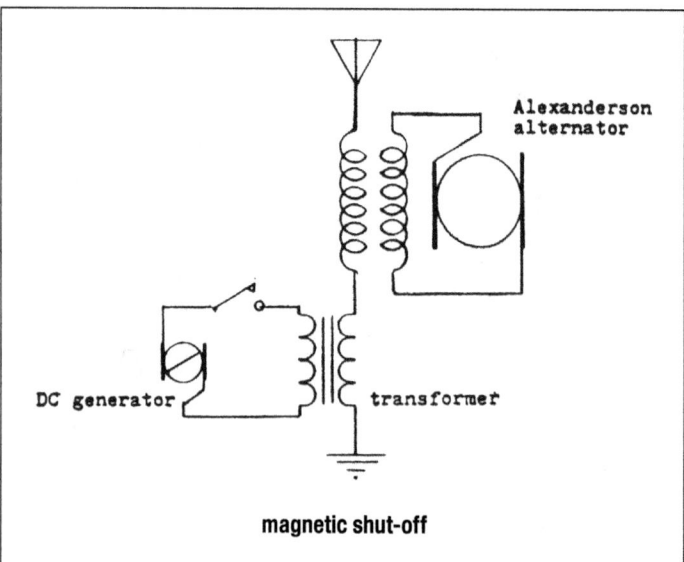

magnetic shut-off

resistance when the current becomes zero, is due probably to the formation of an insulating oxide film on the aluminum." Sounds like semi-conductor talk to me.

The Chaffee gap operated in an atmosphere of alcohol vapor. An airtight spark gap in an atmosphere free of oxygen cannot form the ozone that would corrode its electrodes. Any airtight series gap benefits from a design that permits the introduction of a gas. Various gases can be experimented with, and even air reduced by combustion has been used (including auto exhaust).

As early as 1900 Reginald Fessenden achieved bona-fide radio telephony with a spark transmitter. He probably used a very simple and now-forgotten method known as absorption modulation. A related technology is the magnetic shut-off for telegraphy.

Now we are in the technology of the saturable reactor and the magnetic amplifier. (See the Mag Amp Page at Teslapress for the basics and some detail on the mag amp project shown below.) The mag amp is not Tesla. I have searched his patents in vain for any evidence that Tesla, as sophisticated as he was in magnetics for his motors, was aware of the mag-amp principle. The mag amp is introduced here because it can control the huge energy of Tesla's oscillators. How do you modulate a Tesla transmitter? With a mag amp.

Fessenden in 1906, broadcast a program of music and speech with an 80 kc alternator. Absorption modulation for radio-telephony was installed in the circuits of either antenna or ground. Notice in the illustration that the magnetic shut-off is in the ground circuit, suggesting a respect for grounded radio.

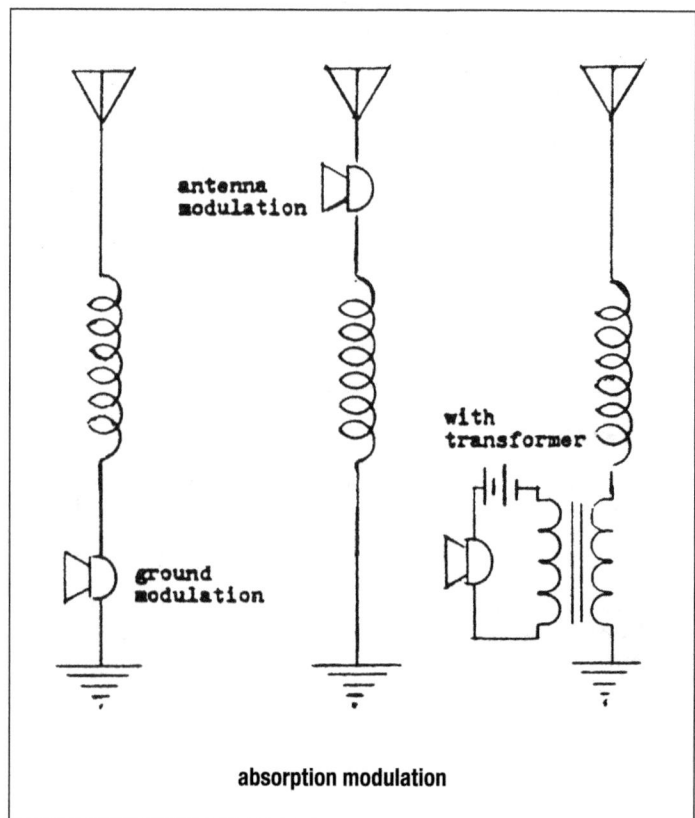

absorption modulation

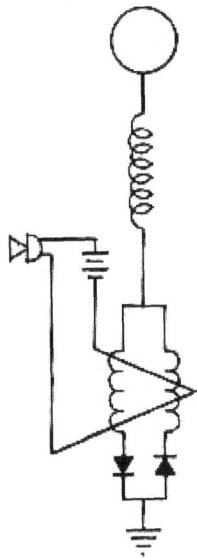

mag amp modulates
Tesla transmitter

Alexanderson alternator schematic

Early alternator transmitters had to be modulated by absorption. Tesla, the inventor of the AC generator or alternator, had experimented with units having many poles and capable of frequencies of up to 20 or 30 kc. These he intended to use not for radio but for his high-frequency lighting.

Tesla abandoned the alternator early on in his experimenting because of its frequency limitations and developed the spark-gap oscillator and Tesla coil. Ironically, it was an alternator quite similar to Tesla's that became the first overseas radio transmitter. Fessenden developed a 50-kilowatt dynamo. Alexanderson scaled this up to 200 kilowatts. Shown is a 2 kw, 100 kc Alexanderson alternator.

These huge alternators (they looked like power-plant dynamos) were put into service in parallel with spark. Alternator radio necessarily remained low-frequency (often so low that voice modulation was not possible) with long antennas and deep groundings. The high-RPM alternator could not be turned on and off for telegraphy and was keyed by a magnetic shut-off.

Tesla said that his transmitters produced continuous waves. Evolved spark transmitters used synchronous rotary gaps, which coordinated the charging rhythm of the capacitor with the frequency. A feature of Tesla's magnifying transmitter design that discourages undamped waves is the single-turn primary. This reduces the induction of the primary circuit so that unwanted oscillation cannot occur there.

The inexpensiveness and simplicity of absorption modulation tempts experimenters who seek a simpler radio technology. I have seen the method described as "crude" in a 1924 radio book where it pops up, but no specifics were given. Is it necessarily low-fi? Does it cause unwanted FM effects, widening the bandwidth unacceptably? Absorption modulators commonly used carbon microphones.

Carbon mics modulated some early transmitters, including broadcast. The overheating of carbon mics could be a problem

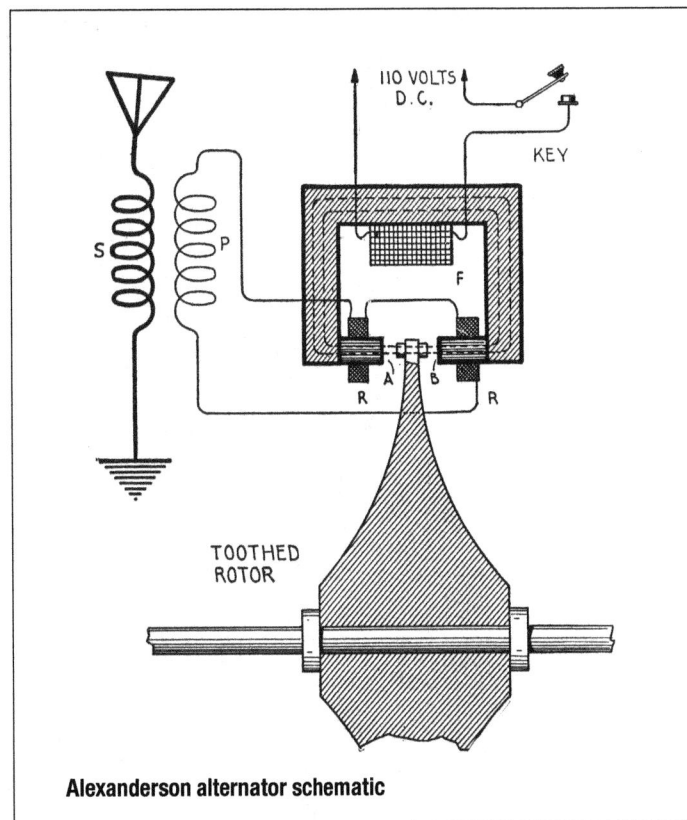

Alexanderson alternator

author's mag amp (160 Kc)

carbon mic
for broadcasting
Kellog, 1920

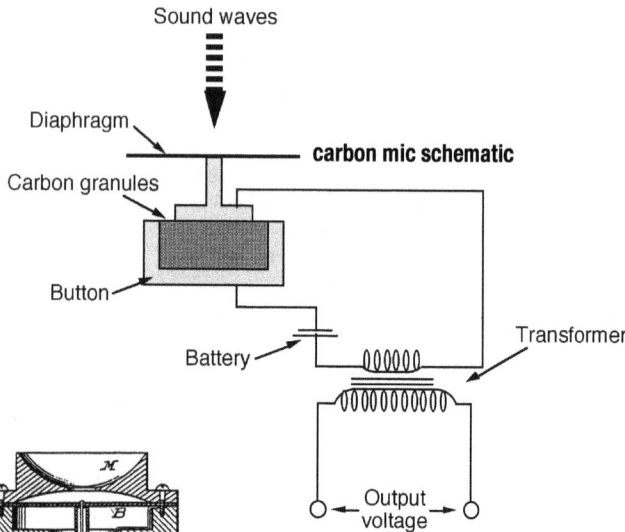

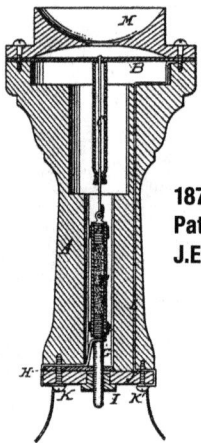

Sound waves

Diaphragm

carbon mic schematic

Carbon granules

Button

Battery

Transformer

Output voltage

1879 carbon mic
Patent 219,544,
J.E. Watson

solid-back
carbon mic

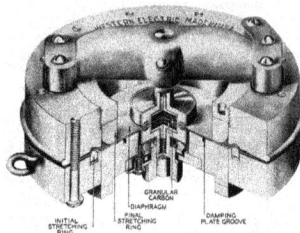

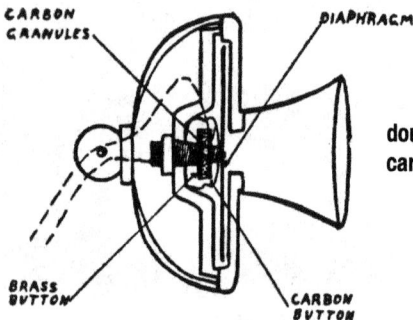

double-button
carbon mic

CARBON GRANULES

DIAPHRAGM

BRASS BUTTON

CARBON BUTTON

in multi-kilowatt rigs. One cure was water-cooling. The carbon mic is a transducer, an amplifier, and a modulator rolled into one simple device. If you wire a carbon mic in series with a battery and a loudspeaker, you have the simplest public-address system. The typical carbon mic had a metal diaphragm attached to a button containing carbon granules. Advanced designs used carbon-impregnated membrane. The carbon-button mic was basic to telephony through most of its history. Some audio engineers of the past believed that the best fidelity was accomplished by the carbon-microphone, particularly in respect to the capturing of subtle emotive values in the human voice.

KQW in San Jose, California broadcast voice and music circa 1912 on an arc transmitter heard for about a thousand miles, and WLW in Chicago boomed on the AM band in the early '30s with a half-megawatt spark transmitter. Both used the antenna-absorption method. Tesla's available notes and patents tell us very little about his particular modulation schemes. His Colorado notes show an "arc controller." But Tesla, the promoter of a world broadcasting system, seemed supremely confident that his magnifying transmitters could be modulated.

The passing of spark marked the end of Tesla-style high-voltage, sudden-pulse transmitting.

CHAPTER 2

Low Frequency

"Oscillations of low frequency," said Tesla in 1919, "are ever so much more effective in transmission, which is inconsistent with the prevailing idea." Radio at low frequencies (below the AM broadcast band) delivers more distance per watt. It is less vulnerable to atmospherics than shortwave. It does not suffer from the fade-in-fade-out syndrome that plagues the shortwave and broadcast-AM bands. Low-frequency radio is less affected by the sun. While much of shortwave is blotted out in daytime, low-frequency propagation is strongest at midday.

The Low End of the Radio Spectrum
(0 – 3.8 MC)

0	200	400	600	800KC	1MC	1.2	1.4	1.6	1.8	2	2.2	2.4	2.6	2.8	3.0	3.2	3.4	3.6	3.8
longwave	AM b'cast								160-m ham			120-m. world b'cast				90-m. world b'cast		80-m. ham	

Longwave 0 – 535kc
(below the broadcast AM band)

0	25	50	75	100	125	150	175	200	225	250	275	300	325	350	375	400	425	450	475	500
ELF NAA, NLK OMEGA open land		LORAN			GWEN LowFER Tesla ------------ world broadcast →				beacons, marine and aeronautic									distress		
	government exclusive		govt. shared		govt. exclusive			govt. shared												

In early radio, the superiority of low-frequency was taken for granted. Shortwave was considered useless. But as radio attracted more practitioners, there inevitably developed a struggle for elbow-room within the limited confines of the recognized radio spectrum.

who owns the ether?

Not surprisingly, the very first entity to claim ownership of the increasingly precious spectrum real estate was Government, in particular that of the U.S. through its navy. The U.S. Navy between 1900 and 1910 made several attempts to put radio under its sole control.

The Navy, which during this period was already developing global marine-communications and radio-navigation systems, was so assured of its right to rule the ether that it campaigned Congress for a monopoly of all of radio, both point-to-point and broadcast, on sea and land world-wide.

The Navy also fought for the phase-out of spark and the development of a more refined radio. The Navy set the specs for industry to develop the new continuous-wave technology.

hams become outlaws

Of course, the spectrum had not been waiting silently for government to step in and regulate it. The ether was already abuzz with activity, much of it conducted by the independent citizen-experimenters, who ultimately became labeled (did they think it condescending?) as "amateurs." The Navy introduced into Congress a series of bills that had the standard formula of providing for different classes of stations that would be registered and licensed, and then making it illegal for any outsider to interfere with these stations. No mention was made of the amateurs, whose transmitting stations (largely spark) were to number about 6000 by 1916.

Thus, in these early legislative attempts, the hams were rendered de-facto outlaws. The Navy's campaign was ultimately joined by the Department of Commerce and by such commercial interests as Marconi's United Wireless. The hams, to defend themselves, organized into the still-active American Radio Relay League (1914).

The legislation that finally passed (1912) did recognize the hams but banished them from any activity in the low frequencies. This was in the spirit of yes-you-can-go-swimming-but-don't-go-near-the-water, for low-frequency radio was all the radio there was. The short waves were then considered not only inferior but unworkable, the desert real estate of the spectrum.

The Government's final solution to the problem of a peoples' experimental radio was achieved, albeit briefly, in World War I, when all amateur radio was flatly outlawed for the duration. The emergency order, which came from the Department of Commerce and was signed by the Navy, banned not only transmitting but short- and long-wave listening as well. The order put radio under the full control of the Navy, which then tried to persuade Congress to make this situation permanent. Ultimately, federal control of radio was to pass out of the control of the Navy and into the Dept. of Commerce, until the Radio Act of 1934 created the FCC.

WWI also provided the opportunity for the government's destruction by dynamite of Tesla's magnifying transmitter tower, which still stood sturdily at Wardencliff, Long Island,

The Electrical Experimenter, 1919

a curiosity to passers-by. The excuse given was that the tower could be used by "spies," but the intention must have been to erase this monument to Tesla's alien radio and wireless power. During WWII the government ordered another blackout of amateur radio.

The hams are still on the defensive. The low frequencies, long under the control of government, have been allocated to military and other bureaucratic functions, to navigation beacons, like Omega (10-14 kc) and LORAN (100 kc), and for weather stations, and time registers. Tesla had suggested low-frequency navigation systems. Some of the military transmitters are humongous, like the Navy's 3000-acre NAA command facility (24 kc) that runs 2 million watts. Then there is ELF (Extremely Low Frequency) down at 76 cycles. ELF uses antennas over 50 miles long.

citizens' radio

The government-military takeover of radio (nearly 100 percent of the low bands and much of the upper spectrum as well) has impacted on the amateurs, who have had to justify themselves as a quasi-government "service" in order to retain the privilege of radio. The amateurs' awkward posture is, "We're an emergency service, but, until an emergency comes up, we'll just be jawboning here to keep these bands open."

The history of the ham is one of experimental adventure compromised by orthocratic entanglement. As early as 1925 the hams were building close functional relationships with the police and military. Many hams are graduates of the Signal Corps. Many are employed by or graduates of the military-industrial complex. In order to stay active during the WWII amateur-radio black-out, some hams enlisted in a military-run civil-defense program called the War Emergency Radio Service. Before government had its own police radio, there was an amateur service that assisted the police in such matters as recovering stolen cars.

An alternative to licensed ham radio is CB and also the lesser-known MURS channels.

The fact that there is any radio that is nongovernment or noncommercial is obscured from the public consciousness. While hams sometimes get a few seconds on the TV news when

CB transceiver
40-channel mobile
(Uniden Pro 510XL, $39)

2-meter transceiver
mobile, 65 watts
(Kenwood TM-281A, $140)

they are performing their public service during an earthquake, the general public is given little awareness via the mass media that such a thing as a citizen's radio exists. On the TV the only character holding a radio microphone is a cop.

The 11-meter (27 MC), 40-channel Citizens Band (CB), still flourishes among truckers and among some ham-like radio enthusiasts who transmit on the upper channels in single-side-band. The five license-free MURS channels reside near the 2-meter ham band and 2-meter equipment can be used. (The MURS channels are 151.82, 151.88, 151.94, 154.57, 154.60, limit two watts.) The license-free low-frequency experimenter band (LowFER) is discussed below.

A colorful wall-poster chart of the entire radio spectrum published by the office of Spectrum Management (a bureaucracy within the Department of Commerce's National Telecommunications and Information Administration) provides elaborate color coding of 28 radio services, including broadcast, radio navigation, amateur, etc., but has no code for a citizen's radio; and the 40-channel citizen's band at 27 MC is ominously absent from the chart.

Regulation of the ether can never be absolute. Some governments have abdicated totalitarian control over broadcasting to independents ("pirates"). Pirate radio and TV flourish where a state radio tries to dominate, and, by the same dynamic, where broadcasting is dominated by a handful of media corporations acting as one, piracy will predictably rise up.

Some people believe in keeping a gun in the house by way of providing for some impending political catastrophe. I'm not a gun person, but I do advocate keeping a transmitter handy. Does the First Amendment guarantee the right to bear transmitters?

The FCC once thought that it controlled CB, but when CB exploded with the advent of inexpensive transceivers (made possible by the phase-lock-loop frequency synthesizer) the government found it could not enlist the cooperation of all operators in a licensing procedure and threw up its hands. CB may be anarchy, but it still works, and the FCC can't do much about those kilowatt linears anyway.

black-box consumers

There is little encouragement for hams or anybody else to explore radio experimentally. The literature testifies to a time when way

opened more easily for the curious and the experimenter. Published between 1900 and 1930 were at least twenty different monthlies for the electrical experimenter, including *The Electrical Experimenter* and *The Radio Experimenter*. These publications were full of projects, including transmitters and receivers, that one could build in the basement workshop from parts easily available at independent dealers. Tesla, who published in the popular press as well as the academic, was a star contributor to some of these magazines.

The reader was assumed to be an experimenter, a builder, and even capable of sustained thought, judging from the prose. Today he is assumed to be, above all, a consumer. The typical ham through the 1930s built his own station from discreet components, transmitter and receiver. By the mid-1940s he was still likely to build his own transmitter (though manufactured units were becoming available), but by then he had been persuaded that the manufactured receiver was superior to anything he himself could build. Many of the radio listeners of yesteryear built their own receiving sets for broadcast-AM listening as well as for shortwave, but by 1930 this became unfashionable due to the abundance of manufactured units.

Since the 1960s the ham has bought both transmitter and receiver. Heathkit and the other popular kits of the 1950s are gone. Today the amateur is encouraged to buy, for $800 and up, both

> ## DO NOT REMOVE COVER.
> ### *NO USER-SERVICABLE PARTS INSIDE.*

transmitter and receiver, boxed into one slick solid-state transceiver that the ham himself cannot repair.

Once an experimenter and builder, the ham of today is encouraged to become just another dependent consumer of black-box electronic components, which he usually cannot repair himself. In the case of broadcasting, the FCC prohibits a licensed operator of an AM, FM, or TV station from building his own transmitter, and the FCC dictates design, hence cost.

With today's integrated circuits, even the do-it-yourself discreet-component builder and experimenter becomes a consumer of tiny black boxes the concealed workings of which he may only dimly understand. Sealing up electronics in black boxes discourages the curious, while it promotes a priesthood. So does the conventional electronics education and its textbooks. The innocent student, from the first day, is snowed under with mathematical formulae, information that could be useful someday if he were allowed to engineer a circuit, but it cannot tell him how electricity works. (Your author in his checkered past was an editor of a series of those "basic" electronics texts. His Tesla work may be an act of penance.) Tesla, a consummate mathematician, was skeptical of mathematical explanations. He called Relativity "a mass of deceptions wrapped in a beautiful mathematical cloak." Tesla had no good words for the electronics theory that grew out of quantum and Einsteinian physics. He is on record as saying that "there is no such thing as an electron."

low-frequency revival

While conventional histories congratulate the amateurs for turning shortwave into a viable radio (the hams invented the shortwave technology that the Navy and others would adopt), the allocation of hams to frequencies above 1500 kc meant their exile from superior turf. The banishment of hams from long-wave along with the romancing of the shortwaves has resulted in the almost complete lapse from public memory of low-frequency radio and what it means. This is particularly true in the US, for there is some long-wave broadcasting abroad, which, incidentally, can be received here to some extent on both coasts.

LowFER

In recent years there has been a revival of longwave among radio experimenters. Low Frequency Experimental Radio people (LowFERs) transmit license-free in the 1750-meter band (160-190 kc) with government sanction under Part 15 of the Federal Communications Act, the same little known part of the law that permits flea-power community broadcasting. The law limits LowFERs to just one watt and to a 50-foot antenna-ground system. Still, under ideal conditions, a legal LowFER signal can be heard for up to 1000 miles, testimony to the power of longwave.

But there are so many competing disturbances in these frequencies that reception is almost impossible in urban environments. Jamming the low band are broadcast harmonics, power-line harmonics, power-line-carried noises from motors and light dimmers and many other inadvertent transmitters of modern electric civilization. The LowFER must escape to quieter rural environments.

LowFERs set up telegraphic beacons, listen for each other, and report reception in their networking websites and newsletters – publications that are in the avantgarde of experimental radio. LowFERs are among the few who build their own. LowFERs remain true to the experimental spirit that marked early radio. They are willing, as one LowFER put it, to "reinvent the wheel."

The LowFER goes mobile into the boondocks where it's electronically quiet, but even there signals may be overwhelmed by GWEN, a nuclear-doomsday Air Force system with a hundred sites across the country. GWEN operates in the same band assigned to the one-watt LowFERs, and its stations rehearse daily, emitting bursts of heavily encrypted radio teletype. The same federal spectrum chart that forgets CB assigns a "government exclusive" code to all of LowFERland.

Most LowFERs are resigned to working experimentally within the challenging constraints of the government-mandated one watt, but some are agitating for more practicable levels of power.

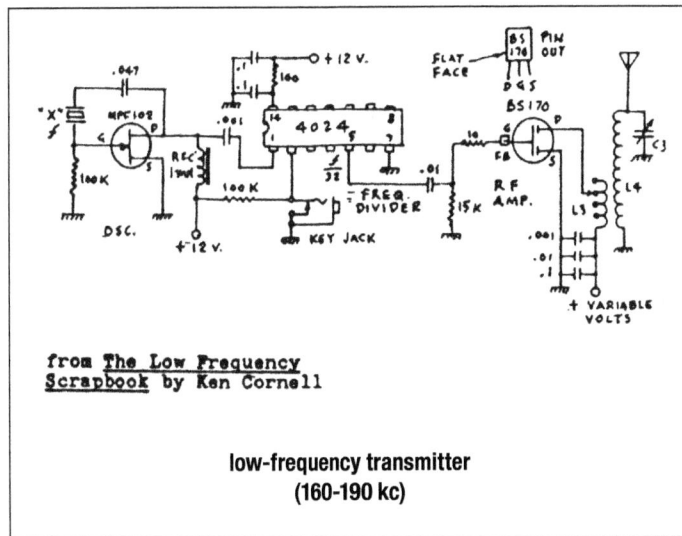

from The Low Frequency Scrapbook by Ken Cornell

low-frequency transmitter (160-190 kc)

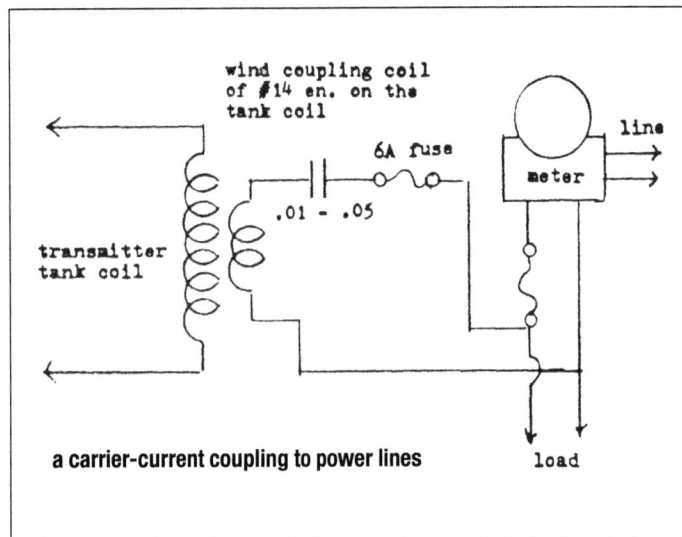

a carrier-current coupling to power lines

Closely allied to the LowFERs are the MedFERs, medium-wave experimenters who operate just above and below the AM band with a 100 milliwatt power limit, the same flea-power limit applied to community broadcasters. The upper MedFER territory has been opened to commercial broadcasters, but these experimenters will continue in the band until displaced.

carrier-current radio

When amateur radio was outlawed in WWII, the only activity open to hams, outside of enlisting in the War-time Radio Service, was a low-frequency radio propagated over the power lines. Called carrier-current, or wired wireless, this little known mode of radio is particularly effective at low frequencies. Line transformers tend to choke out high frequencies, but low frequencies conduct through. The WWII hams worked voice and code down at 160 to 200 kc.

Power companies use a low-frequency carrier-current system to transmit information between stations. Power-line carriers are another nuisance to low-frequency listeners.

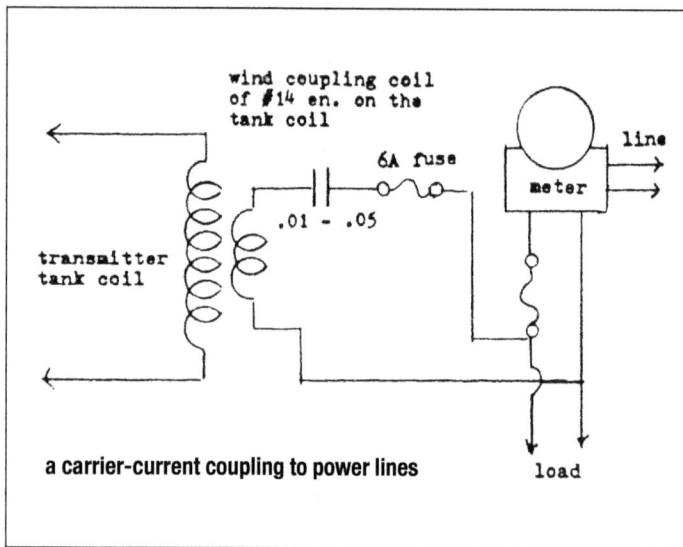

a carrier-current coupling to power lines

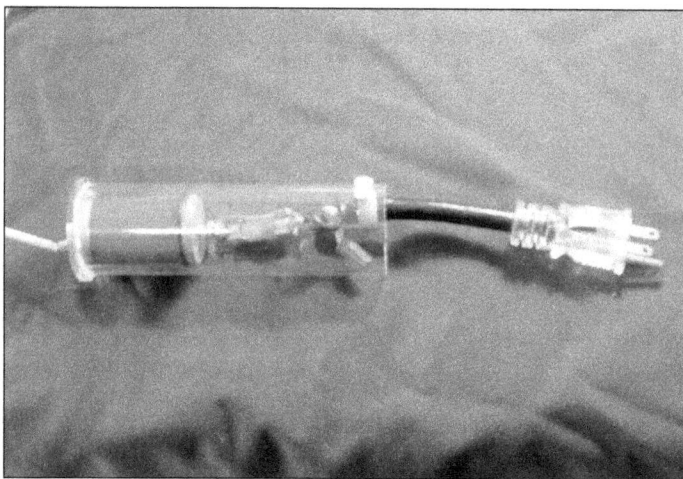

author's plug-in isolator for carrier-current broadcasting

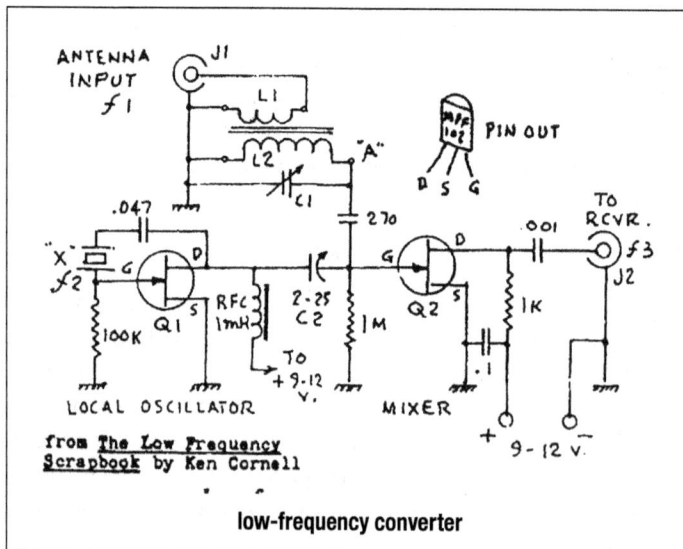

low-frequency converter

AM-band transmitting over the power lines (carrier current) is an option for the independent. The plug-in isolator shown is from my how-to "Enhancing the Ramsey AM-25" posted on Teslapress. A high-voltage capacitor accomplishes isolation. A safety fuse is put in series. The isolator can be wired for the hot or the neutral line. Receivers need not be coupled directly to the lines. Even car radios may get signal from the overhead wires.

The AM band is five times higher in frequency than the low frequencies a power company or a WWII ham could use, but AM-band broadcasting is possible over the power lines. The lower end of the band is preferable. My college AM carrier-current station (WRCU, Colgate) did a good job of getting around the entire campus as well as much of the adjoining small town.

open land

Did you know that in the basement of the very crowded, highly controlled radio spectrum, down at 0-9000 cycles, there is no government regulation? Thus, this unallocated very low frequency band is open to unlicensed experimenters at any power level. Conceivably, one could set up point-to-point communications using just everyday audio amplifiers. An underwater diver communication system uses 10-watt audio amplifiers. Output is connected to widely spaced submerged electrodes as antennas. The idea might be tempting to audiophiles who would be radiophiles. A problem would be the noises, both civilized and natural, that abound in this band, particularly the 60-cycle power-line hum and its harmonics. Another problem is that a single voice-modulated signal would consume the entire band within its range for the duration of the transmission.

low-frequency listening

When shortwave came into fashion, the lower frequencies became de-facto off-limits to the listening public. But lately, along with the LowFERs, the low-frequency listener is coming back. There is a growing curiosity about what goes on in the forbidden territory below the AM broadcast band.

To tune in reliably to distant LowFER beacons takes a sophisticated receiver with crystal filters and one-kilocycle dial resolution, but I can testify to the enjoyment provided by seeking stronger beacons on my simple, one-transistor, uncalibrated home-built regenerative unit shown in Part III, Tesla DIY. If you own a shortwave receiver, you may want to try a solid-state low-frequency converter like the one shown above. Some commercial multi-band receivers are built to tune the low band.

What you seek to tune in are low-frequency non-directional beacons. These are radio-telegraphic navigation "lighthouses" that repeat slow alphabetic identifiers (good code practice). Called NDBs in LowFER jargon, marine non-directional beacons run only about 5 watts, while the FAA's aeronautic beacons can go to a kilowatt. There are directories that can tell you the location of beacons heard, if you want to take it that far. Also you may hear maritime ship-to-shore telegraph as well as big military transmitters. Receiving the megawatt foreign low-frequency broadcast stations is a possibility if you live on either U.S. coast. Any low-frequency listening is subject to suppression by civilized electrical noise and benefits from a non-urban environment. It is in urban listening that the expensive well-filtered receiver pays off.

natural radio

With a very low-frequency (VLF) receiver (below 30 kc), you can hear the sounds of natural radio: spherics, hisses, pops, chirps, tweaks, whistlers, a phenomenon called the dawn chorus, and some other signals you just have to wonder about. Perhaps this is where to listen for interplanetary signals (contrary to the high-frequency assumptions of SETI).

Experimenters today who tune in to nature's radio use receivers of various degrees of sophistication, but, according to Michael Mideke in his *Sounds of Natural Radio*, "given a large enough antenna, even amplification can be dispensed with; signals can be heard using nothing more than headphones connected between antenna and ground."

signals from Mars

In 1899 at Colorado Springs, Tesla listened to the sounds of low-frequency radio and conjectured that, among the natural disturbances, he was hearing signals from an extraterrestrial civilization, perhaps Mars. "A clear suggestion of number and order," he noted; "impossible to think accidental. A purpose was behind these signals." Appreciate that Tesla tuned in on a primitive ether which was free from man-made signaling of any kind. A contemporary outer-space listening project called Search for Extraterrestrial Intelligence (SETI) expends most of its energy sorting out the abundant noise of earth-bound emissions.

Tesla would get a laugh out of SETI's hypothesis of propagation, which he would see as another Hertzian perversity. SETI assumes that the super-intelligent outer-space communicator would choose a UHF frequency compatible with the resonant frequency of hydrogen. Propagation is via hydrogen atoms floating in an empty space, say the quantum-fashionable SETI theorists, who believe the ether to be long-ago disproved. But Tesla was listening at low frequencies near the resonant frequency of the earth itself. His hypothetical communicator from space would have the intelligence to transmit to other planets at their natural resonant frequency of electrical vibration, a function of that planet's diameter and composition.

Colorado Springs lab

CHAPTER 3

Conduction through the Ground

The propagation of radio signals and wireless power Tesla saw as a matter of conduction. Tesla's radio-conductive path was not a circuit in the conventional sense but was modeled on the single-wire-without-return principle that he demonstrated in his high-frequency, high-voltage lighting system. Electric researcher Eric Dollard calls this one-way conductive effect "longitudinal electricity."

Tesla's radio can conduct through the earth or through the sky. Tesla said, "The earth behaves simply as an ordinary conductor." He said the earth was superior conductor; but, pertinent to sky transmission, Tesla is on record in a patent as saying that "transmission through elevated strata encounters possibly less resistance than copper wire."

Conventional theory never hints at radio propagation being anything like conduction but represents it as some kind of radiation. This is more than a textbook convention; a whole science of antenna design is based on this assumption, which Tesla dismissed.

an etheric medium

What is radio, anyway? The texts, like my *Radio Amateurs Handbook*, speak of radio as "'radiation." But, at the same time, it is a "wave." Then again, in the same paragraph of the *Handbook*, which is just echoing respectable physics, we are told that it is "traveling electrostatic and electromagnetic fields." Whatever it is, some sort of almost material emanation is presumed to be issuing from the antenna into a void.

Tesla, however, saw radio as a "disturbance of the medium," or a "commotion in the medium." He writes of "transmitting an electrical movement to the environing medium." As we have seen in Part I, it was an article of faith in Tesla's time that there existed a unified field that permeated all being, all matter, including solids, liquids, and even what is called the vacuum of space. This pervasive "ether," as it was called, is the medium of radio. The transmitter sets up a disturbance that produces, says Tesla, "alternate compressions and rarefactions in the medium." This suggests some kind of elastic continuum that conducts standing waves.

Coming from this view, all electric phenomena – charge, polarity, oscillation – involve some kind of strain or vibratory disturbance of the equilibrium of the ether. The etheric perception of radio came naturally to early experimenters. Loomis writes of his aerial telegraphy as "causing electrical vibrations or

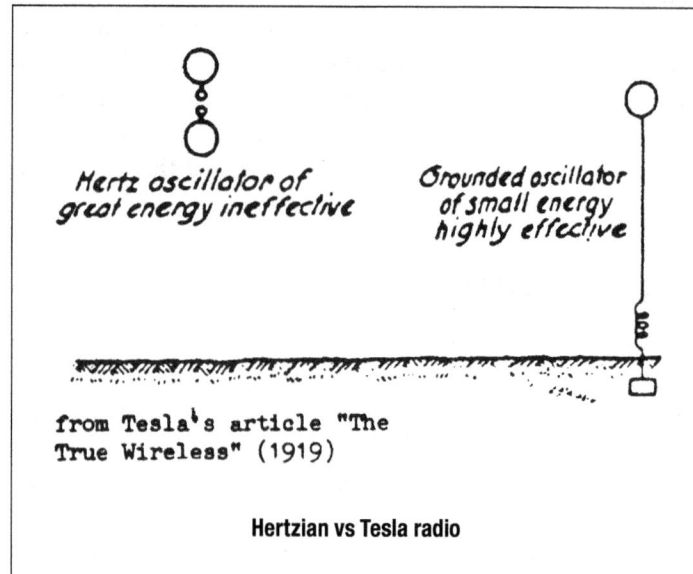

from Tesla's article "The True Wireless" (1919)

Hertzian vs Tesla radio

waves to pass around the world, as upon the surface of some quiet lake one wave circle follows another from the point of disturbance to the farthest shores."

In Tesla's radio, effective transmitting is achieved by setting this elastic medium vibrating with a sudden high-voltage whack, like that created by the discharge of a capacitor through a spark gap.

grounded radio

Hertzian theory encourages us to think in terms of aerial radio: "the air waves," "on the air." Extreme importance is placed on the antenna and its configurations, which can become byzantine. But Tesla's radio is not aerial. It is ground-conduction. The lower end of the energized coil is rooted in the earth. Pure Hertzian radio has no such natural load. Modern radio regards earth ground as an electronic toxic-waste dump to which noise is conducted. Tesla said radio "should be designed with due regard to the physical properties of the planet and the electrical conditions obtaining in same."

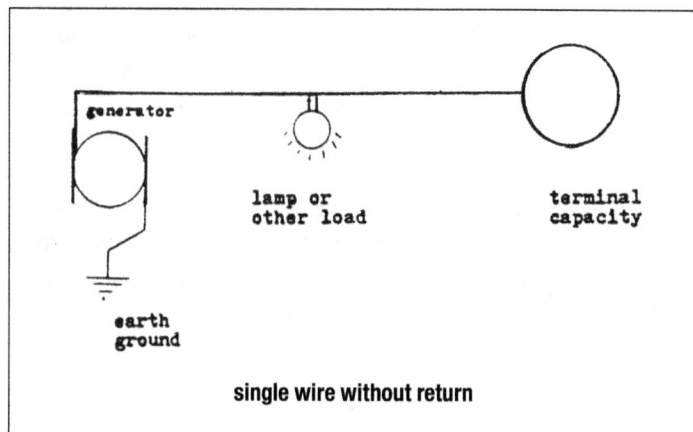

single wire without return

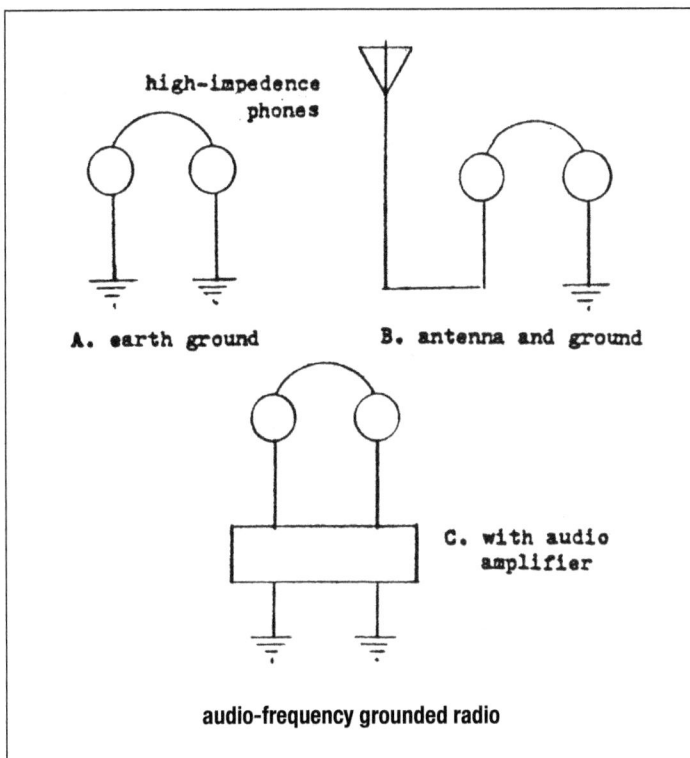

audio-frequency grounded radio

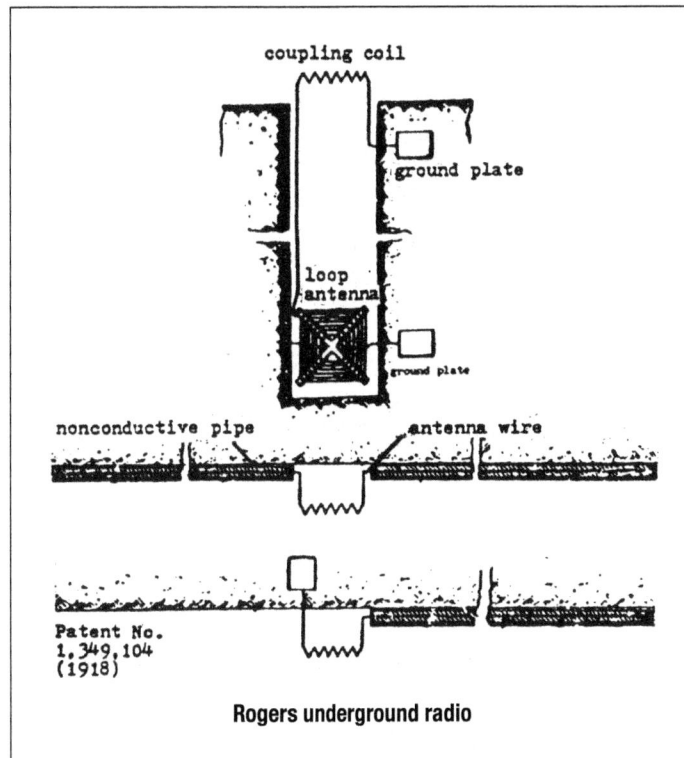

Rogers underground radio

The electrical vibrations of the transmitter are communicated to the ground," says Tesla. Here they set up standing waves, "outgoing crests and hollows in parallel circles." He said, "The terrestrial conductor is thrown into resonance with the oscillations impressed on it like a wire." Tesla was convinced that the waves were not "electromagnetic" since such waves were not likely to travel through the earth. Tesla's ground wave, unlike the Hertz aerial wave, does not travel uniformly at the speed of light and can even reach a velocity that is infinite.

Oscillating the entire electrostatic Earth was among Tesla's many ambitions, and he observed that "the planet behaves like a perfectly smooth and polished conductor of inappreciable resistance." He went on to characterize the globe as having "capacity and self-induction uniformly distributed along the axis of symmetry of wave propagation." In the same text (a patent applied for in 1902), Tesla added that these standing waves propagate "without attenuation." That is, they do not diminish in intensity over distance.

no ionosphere

Tesla believed Hertzian aerial radio was a waste of energy. Tesla suggests that there is some radiation component in a radio transmission but that it is a weak and incidental output. He thought the theory that high-frequency signals traveled long distances by bouncing off some high-altitude, radio-reflective layer called the ionosphere was "an utter impossibility."

Tesla was not alone in disputing the existence of the Heavyside Layer, as the ionosphere was called then; it was quite controversial for a time before it was frozen into official science. Tesla said that Hertz wave theory "by its fascinating hold upon

the imagination, has stifled creative effort in the wireless art and retarded it for 25 years."

inadvertent grounded radio

In World War I, when field telephones were used that stretched long wires over the surface of the ground, it was discovered that one could eavesdrop on the conversations of opposing forces simply by connecting headphones to ground rods. This discovered, effective monitoring systems were developed by both sides that consisted of two widely separated ground rods connected to a sensitive audio amplifier. Here was a Tesla-like radio: low-frequency conduction through the earth.

There is no reason you cannot use this principle for an audio-frequency radio in the unregulated open land below 10 kc. When you consider that radio can happen in the audio band, you have to wonder about the term "radio-frequency"(rf), which seems to imply that radio is a function of frequency.

Disturbances caused by the various electrical happenings in nature, such as lightning-created static crashes, an occasional whistler, and other strange sounds can also be heard using the above system, and the WWI eavesdropping apparatus was sometimes totally jammed by such natural activity.

underground radio

A follower of Tesla, James Harris Rogers, patented (1920) a radio system in which both sending and receiving antennas were sunk completely underground or underwater. Rogers had his own doubts about Hertzian radio. He wondered, "If 50 units of

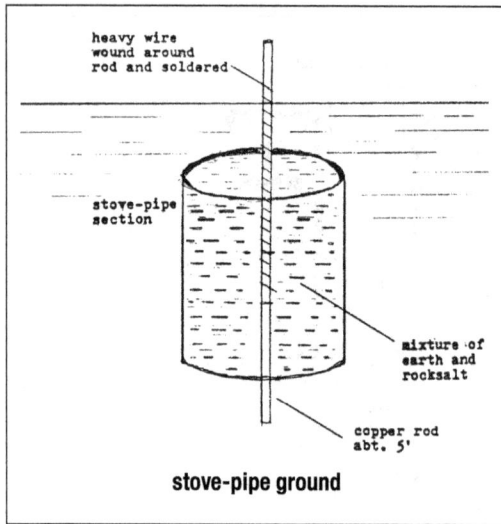

stove-pipe ground

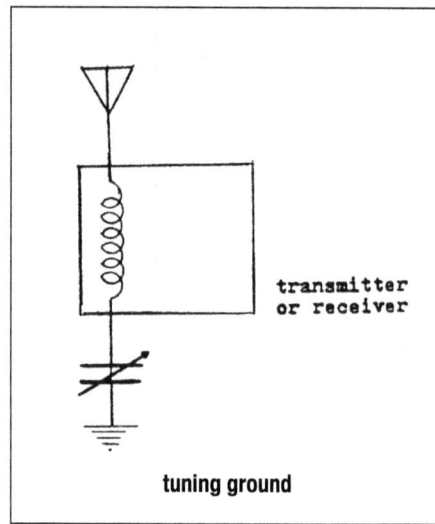

tuning ground

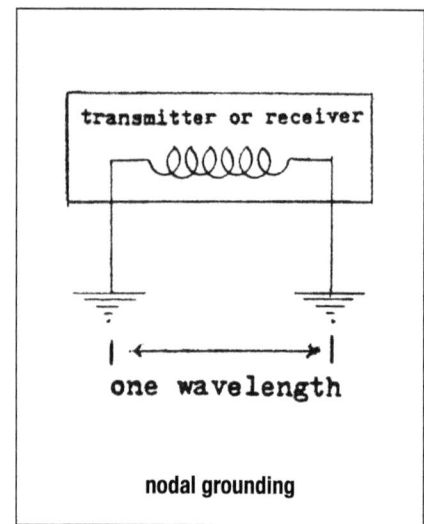

nodal grounding

power are passed into the aerial, then what becomes of the equal amount of energy which passes into the ground?"

Roger's was working at low frequencies, but nothing in his patents limits his grounded radio to the low band. The Rogers underground achieved superior signal strength. According to accounts, reception was free from atmospheric static, and there was no diminishing of the signal in daytime. The Navy used the Rogers system secretly in WWI and announced it to the public in 1919.

Tesla said, "The great amount of energy which can be conveyed to a (receiving) circuit by conduction through the ground, makes it appear possible that the necessity of elevating terminals may be dispensed with."

The famous editor Hugo Gernsbach, who published much on Tesla in his time, said, "All our pet theories on wireless are thrown into chaos," and he predicted "a war to the knife between our wave-propagation theorists and the new school of ground-impulse savants." Gernsbach's editorial in the March 1919 issue of the Electrical Experimenter predicted that "the greatest pride of the radio amateur, the aerial on top of his house … is doomed … As for commercial stations their towers are doomed shortly for the scrap heap."

This noted, the Rogers underground vanished from the media, and Rogers' career took a downturn reminiscent of Tesla's.

Stubblefield

In 1902 Nathan Stubblefield demonstrated a grounded wire-less-telephone system of his own invention. Sticking iron rods into the ground at the Virginia shore of the Potomac River, Stubblefield communicated with a steamship a half mile away. Stubblefield's grounded radio was powered by the high-frequency output of

earth battery

his earth battery (patent No. 600,457), a type of space-energy receiver, which shows how the occult science of free energy nudges up against that of radio, another instance being the free-energy work of T. Henry Moray. Loomis, too, used a power supply that obtained electricity from the atmosphere.

grounding technology

"The ground should be made with great care with the object of reducing its resistance," said Tesla. To ground his Colorado magnifying transmitter, Tesla buried a 20 by 20-inch copper screen 12 feet down in the arid soil. Over the top of the screen he spread a layer of coke. He flowed water over the spot continuously. Beneath Tesla's Wardencliff tower, a shaft descended 120 feet into the earth. Out from the bottom of the 12-by-12-foot shaft, side tunnels extended radially. They were carbon-blackened to enhance conduction. Situated near the Long Island shore, the giant transmitter was thus grounded to the oceans.

Such solid grounding has become another lost radio art. Until the 1950's commercial broadcast receivers had a ground terminal on the chassis that one was encouraged to connect to a cold-water pipe or other ground. Some houses, like the one I grew up in, had a radio ground wired into a wall socket.

A 1930s shortwave magazine recommended a stovepipe ground, a buried stove-pipe section filled with a mixture of soil and rock salt, which attracts moisture. Charcoal can be mixed in. (Tesla used coke.) Down the center runs five feet of solid copper rod or galvanized pipe around which is wrapped a heavy connecting wire. Any conductor used to connect a radio to ground should be as short as possible and made with the heaviest conductor, such that an excessively long run of say, twenty feet is best done with a wire gauge of zero or bigger or maybe with copper tubing or galvanized pipe.

Clamping securely on to cold-water plumbing near to its ground-entry point is good grounding, but this can be supplemented by other grounds, like earth rods or connections to existing fence posts. Any body of water makes a good ground. Connect to a large submerged metal object. Any radio transmitter or receiver becomes hyperactive near or connected to a body of water.

Older schematics show a variable capacitor between output coil and ground, comparable to an antenna tuner. I've adopted this practice of tuning ground with good results, including noise control in receiving.

Tesla's radio assumes standing earth waves, and Tesla recommends grounding the two ends of the receiving coil one-half wavelength apart at a wave's nodal points, the location of which one determines experimentally. This arrangement is particularly important for reception of wireless power. Having established a solid ground, some interesting experiments can be conducted in receiving. Disconnect the antenna. Does the signal vanish? What happens when you connect ground to a receiver's sacred antenna terminal? Will the Hertz police arrest you if you try any of this?

CHAPTER 4

Resonance

The idea of "properly harmonized coils and capacitors vibrating with sympathetic reinforcement and thereby magnifying effects by tremendous ratios is at the heart of Tesla's technology. Tesla's basic radio tuning "tank" circuit (coil plus capacitor between aerial and ground) is, all by itself, a powerful resonant signal amplifier and a beautifully simple one. Tesla suggests the power of a coil to vorticize energy. But as radio developed over the years, the tank coils in both transmitters and receivers shrank in size, and the result was a loss in gain that was compensated for by the addition of stage after stage of complex amplification circuitry using vacuum tubes.

Tesla watched this development with bewilderment. For transmitting, the resonantly tuned circuits of the Tesla coil provided the high electromotive force Tesla thought necessary for long-distance radio-conduction. For receiving, the big loading and tuning coils "magnified effects," as Tesla liked to put it. So why depend on complex, multi-stage circuitry to amplify? It didn't make sense to Tesla. He said, "My plans involved the use of a highly effective and efficient transmitter conveying, at whatever distance, a relatively large amount of energy. The receiver itself is a device of elementary simplicity ... In such a system resonant amplification is the only one necessary."

Taking the tank circuit another step out, Tesla conceived of a receiving coil made of glass tubing that was filled with a rarefied gas, thus offering almost no impedance to the signal. This super-conductivity would, he reasoned, result in tremendous gain.

I've heard tell of a radio experimenter in Portland, Oregon who built the basic crystal set on an obscene scale with a broad-cast AM-band tank coil four feet in diameter and six feet long, wound of copper tubing. Tuned with a big old commercial variable capacitor, the single-stage set, void of amplifier stages, drove an eight-inch loudspeaker with room-filling volume.

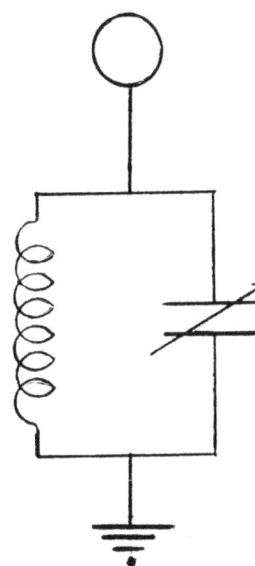

earth resonance

Tesla calculated the frequencies and pulsations of his trans-mitters with an eye to resonating the earth. Earth resonance is fundamental to Tesla's grounded radio and wireless power. Tesla suggested that the electrostatic earth resonated at a particular frequency, and seems to be suggesting that the closer the vibra-tion is to that frequency the greater the magnification of effects.

What is this magical earth-resonant frequency? In his Colorado notes Tesla says the wavelength is 5200 feet or 1737.7 meters, which works out to about 173 kc. Tesla's Colorado transmitter ranged from 60 kc to 190 kc; 170 kc was a typical operating frequency, according to the Notes. In a 1920 patent, however, Tesla specifies that the frequency should be "smaller than 20 kc, though shorter waves may be practicable." The lowest he would allow would be 6 cycles per second, to which he adds that "paradoxical as it may seem, the effect will be in a region diametrically opposite the transmitter." In another fundamental radio patent, Tesla says he has learned to set up standing waves in

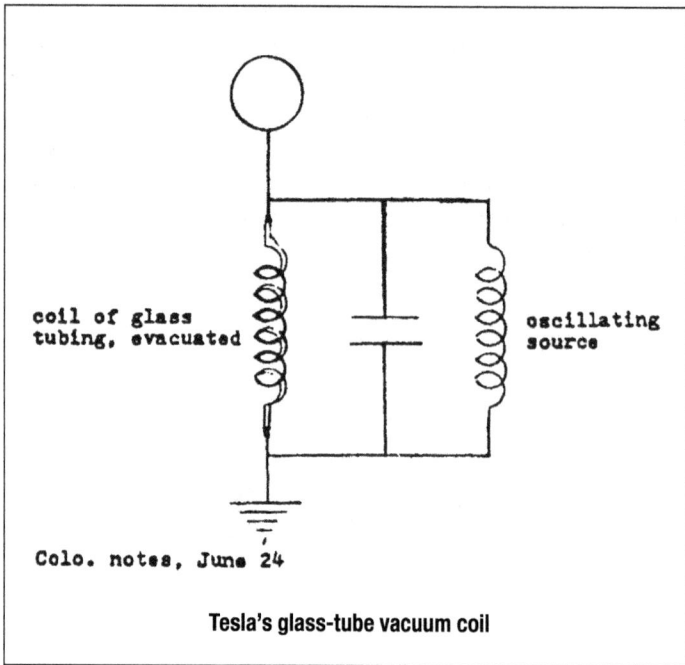

coil of glass
tubing, evacuated

oscillating
source

Colo. notes, June 24

Tesla's glass-tube vacuum coil

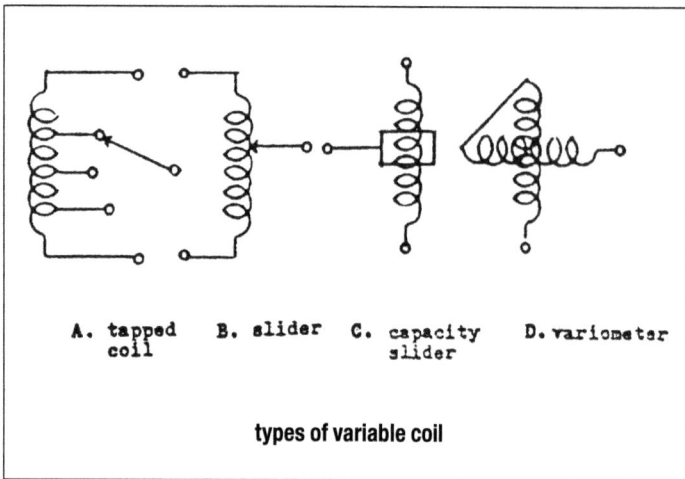

A. tapped   B. slider   C. capacity   D. variometer
   coil   slider

types of variable coil

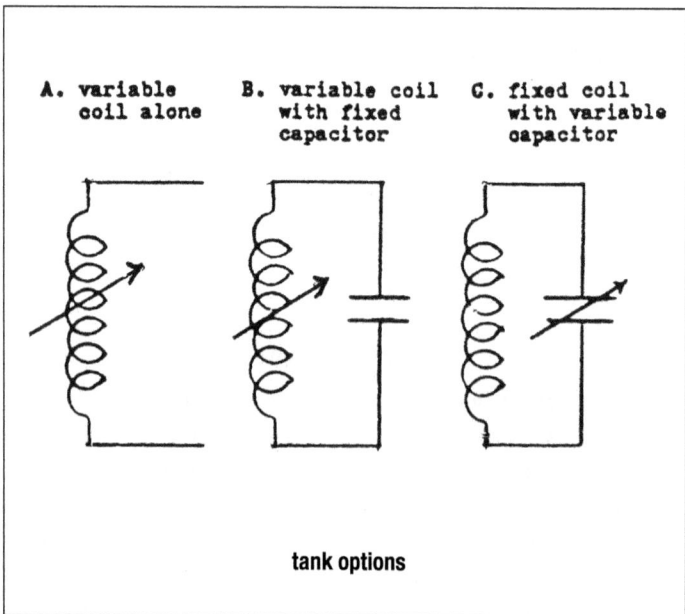

A. variable   B. variable coil   C. fixed coil
   coil alone   with fixed   with variable
                     capacitor   capacitor

tank options

the earth. "I found," he says, "their length to vary approximately from 25 to 70 kilometers." This works out to 428 cycles to 12 kilocycles. If earth resonance happened for Tesla below 60 kc, then he did not achieve it on his Colorado transmitter unless it was by harmonics.

In the 1920 patent Tesla said, "The most essential requirement is that, irrespective of frequency, the wave or wave-train should continue for a certain interval of time, which I have estimated to be not less than 1/2 or probably .08484 of a second — the time taken in passing to and returning from the opposite pole at a mean velocity of about 471,240 kilometers per second."

Fashionable thinking on the earth-resonance mystery hypothesizes a waveguide between the earth and an ionosphere. The Schuman Cavity, as this waveguide is called after its theorist, is said to resonate at exactly 7.83 cycles. The Schuman cavity is a Hertzian notion that would set Tesla spinning in his grave. Tesla said the idea of a radio-reflective ionosphere is a "fiction." He spoke only of resonating the earth itself.

the resonant tank

Of his Colorado transmitter, Tesla wrote, "The vibrating system is formed by a continuously variable and exactly determinable inductance and a capacity standard, or by an inductance standard and a continuously adjustable condenser, or by a system in which both these elements are continuously adjustable." Modern radio technology has settled on the second of these options for tuning — fixed coil and variable capacitor. This is the arrangement since the 1930s for both transmitters and receivers. But early radio shows a mix of options to accomplish tuning. Sometimes there is no capacitor, and tuning is accomplished with a variable inductance alone. Variable coils include the tapped coil, the slider, and the variometer, in which one coil rotates within another. (See Part 3, Tesla DIY for some information on building radio tank circuits.)

The tradition of the tapped and slider coils is preserved among builders of the crystal set. Almost forgotten is the capacity slider, a section of tubing which encircles the coil except for a narrow slit. There is a revival among LowFERs of the elegant variometer.

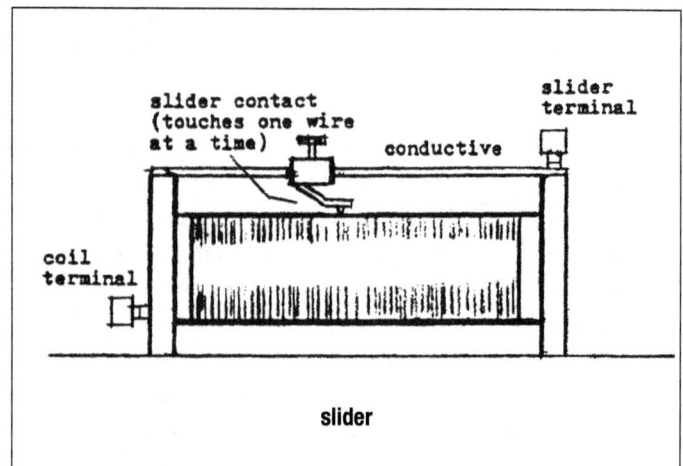

slider contact
(touches one wire
at a time)

slider
terminal

conductive

coil
terminal

slider

World System

When the magnifying transmitter tower at Wardencliff was still under construction (1902), Tesla published a brochure advertising the project as a prototype for a global radio communications corporation. Called the World System in the brochure, it would serve as a multi-frequency wireless traffic center for all existing telephone, telegraph, and stock-ticker services around the planet. Built on Tesla's earth-resonant radio principles, the System would also carry a universal time register, navigation beacons, and even facsimile transmissions. It would also serve as a global system of broadcasting. Tesla was among the first to suggest broadcasting of news and entertainment to the public; only point-to-point signaling had been experimented with up to then.

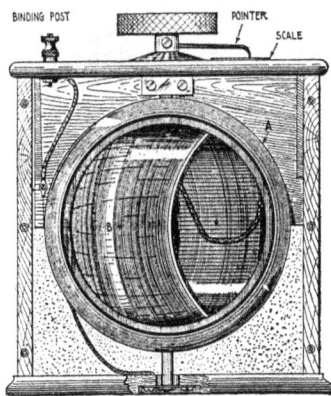

variometer

Tesla's tower was never completed beyond the rugged framing, and the World System fantasy collapsed when Tesla's backing from J. P. Morgan was pulled suddenly from under him. This crushing event signaled the end of Tesla's official career.

The system Tesla describes represents a huge jump forward from any of his radio technology documented in his Colorado notes, his patents, or anywhere else. The brochure suggests the achievement of precisely tunable, high-Q, limitlessly powered transmitting of multiple channels from a single point, transmitting that could be voice-modulated as well.

Little technical detail is given, but the promotional literature suggests that Tesla may have been planning to rely heavily on multiplex techniques to cut through noise. "My individualized system with transmitters emitting a wave complex and receivers comprising separate tuned elements cooperatively associated." He called the technique (actually an AND gate) a "combination lock" and boasted that any degree of safety against statics or other kinds of disturbance can be obtained." The receiver is so designed that it responds "only through the joint action of the tuned elements."

Was Tesla a fascist?

You have to wonder since his World System would have taken radio right off the bat into global centralization, a fantasy of control beyond the dreams even of the U.S. Navy. As it turned out, a quarter century after Tesla proposed his System, broadcast radio, particularly within the U.S. was still quite diversified with hundreds of AM stations, most running only 100 to 500 watts. National networks were still undeveloped. Stations were owned by entrepreneurs, local newspapers, colleges, churches, retailers.

Global radio ultimately became the BBC and VOA. Even today, when multinational monopoly media has put into place

a decadent Hertzian equivalent of Tesla's World System using satellites, including direct-satellite TV, the broadcasting of signals across national boundaries is hotly resisted at diplomatic and other levels, although this battle gets no coverage in the mainstream media.

Tesla saw his World System as a civilizing force: "It will be very efficient in enlightening the masses, particularly in still uncivilized countries and less accessible regions." What did Tesla mean by civilization? He said, "No community can exist and prosper without rigid discipline." He said, "Law and order absolutely require the maintenance of organized force." Tesla said government "should prevent the breeding of the unfit by sterilization; and the deliberate guidance of the mating instinct."

Tesla is admired for his technological purism, his insistence that a machine possibility be carried to its logical conclusion. Society, too, was a machine, and it needed perfecting. The World System fitted into Tesla's ideals of social order. His logical conclusion for organizing radio was a system that was centralized, omnipotent, and global.

Wardencliff

CHAPTER 5

Sensitive Device

Tesla gave a lot of attention to the development of the ideal apparatus for detecting disturbances in the medium. Tesla let his imagination run free in his quest for the optimal "sensitive device," as he called it.

the coherer

Tesla's detector research was paralleled by many others at the time. The popular detector among radio experimenters was the coherer. This is simply a glass tube partly filled with small metal chips or filings. Strained to near-conduction by battery voltage, this early semiconductor mysteriously switches on when an oscillating disturbance is present. A tap is needed to reset the coherer back to non-conduction. Breaking the battery circuit also works. Tesla improved the coherer by setting it into constant rotation at about 16 RPM so it would automatically reset. Changing the rate of rotation controls sensitivity. In a patent he also mentions a vibrating coherer. Constant motion suspends the coherer's chips in space, making them more susceptible to disturbances.

The coherer's internal chips or filings are ideally of uniform size. They are cleaned thoroughly in alcohol. Chips are ideally a mixture of nickel and silver, but other conductive materials can be used. Ideally, the chamber is evacuated, but not necessarily.

Until 1902 the coherer was the only detector in wide use. It dropped out of use about 1912. Tesla must have regarded the

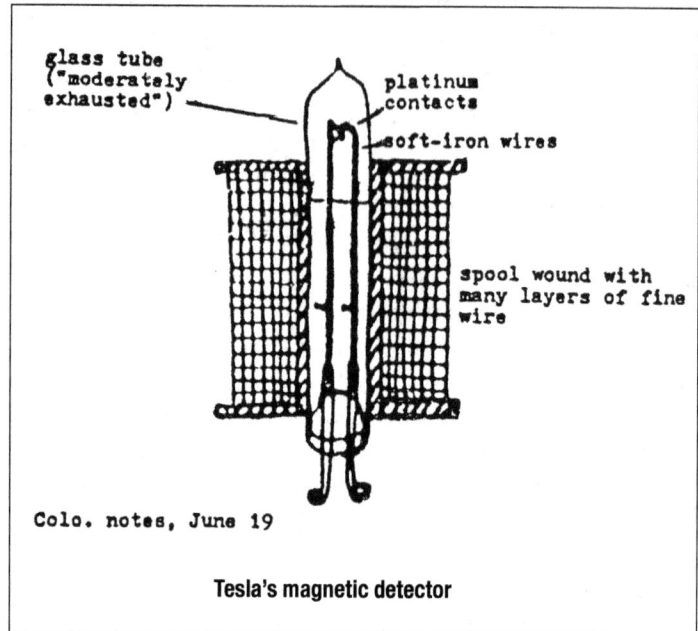

glass tube ("moderately exhausted") — platinum contacts — soft-iron wires — spool wound with many layers of fine wire

Colo. notes, June 19

Tesla's magnetic detector

coherer as a passably reliable sensitive device because he used it in his robot boat (patented 1898) and in his Colorado lightning-tracking experiments (1900).

Tesla's detectors

Tesla explored the possibilities of many other sensitive devices. His patents show a rotating rectifier, a precursor to such static rectifying diodes as the crystal detector. Tesla replaced his rotating rectifier with a vacuum-tube diode. This is mentioned in the Colorado notes, but there is no patent. I've seen mention of Tesla exhibiting, at the 1893 Chicago World's Fair, a vacuum-tube receiver for voice and music. This must have used for its detector a vacuum diode.

Tesla applied the idea of straining a device to near-conduction, as in the coherer, to a diode vacuum tube and to semiconductors having thin-film dielectrics. Another Tesla detector, limited to telegraphy, was visual. It used the deposit of a thin film on a glass surface. An iodine solution, upon being stimulated by a radio disturbance, releases a conductive haloid

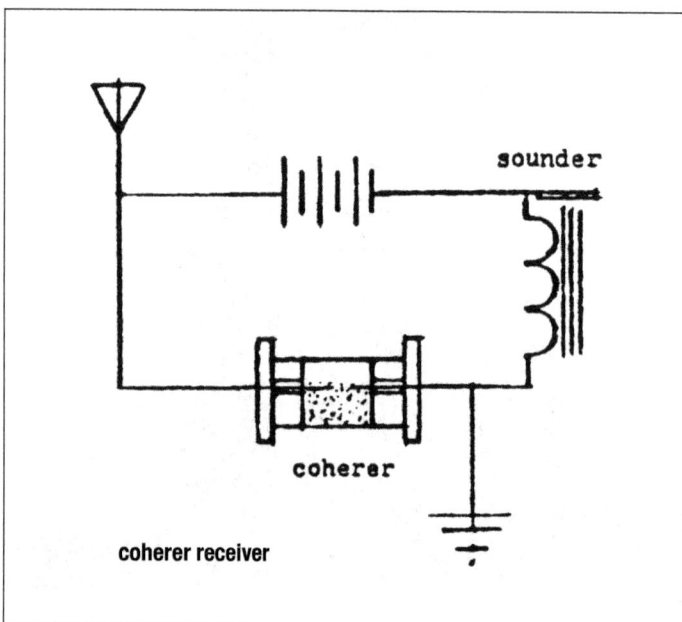

sounder

coherer

coherer receiver

coherer

film onto the glass screen. Battery current, conducted through the film, destroys it, thus erasing the screen. A sort of telegraphic TV, this is a precursor to the liquid-crystal display.

Tesla's Colorado notes show a magnetic detector (See chapter opener). It consists of a coil of many hundreds of turns of fine wire around glass-enclosed wires of soft iron which are magnetically stimulated by a disturbance. Marconi used a "magnetic detector" that worked on another principle. A pair of horseshoe magnets slowly revolved over an electromagnet the windings of which were connected to earphones. This detector was the immediate successor to the coherer.

the rotating-brush detector

Tesla noticed that the megavolt streamers from his Colorado coils were extremely responsive to the slightest changes in etheric conditions and wondered how this phenomenon might be applied to the sensitive device. Later, working with high-voltage, high-frequency currents in vacuum, he discovered the "rotating brush." This is an eerie emanation, a high-voltage brush discharge, from a spherical conductor exactly centered in an evacuated glass bulb. The device resembles the familiar plasma globe, but these contain gases. The result is different in a vacuum.

The brush resolves into a rotating ray so sensitive that if you approach it from a few paces it will turn away from you. Tesla found that a small one-inch magnet "will affect it visibly at a distance of two meters, slowing down or accelerating the rotation according to how the magnet is held relative to the brush."

There is no information on how Tesla planned to harness the rotating brush as a sensitive device, but he wrote that it was "undoubtedly the most delicate wireless detector known."

One associates high voltages with radio transmitting, but the rotating brush brings high voltage to radio detection. Tesla believed the ether to be at high tension, and believed that the best way to engage the ether is at high tension. This is why the centered glass-enclosed conductor in the rotating brush is at high voltage. This is also why Tesla provided the spherical aerial capacity on his magnifying transmitter with a thick, dielectric coating of rubber.

Tesla's receivers

Over 50 different receiver circuits are to be found in Tesla's Colorado notes. They show various configurations of sensitive device, tuning coil, capacitor, rotary interrupter, and battery source. If early radio engineers had studied these notes (not published until 1978), they would have found circuits that lay the foundation for two receivers which would dominate the history of radio: the regenerative, and the heterodyne.

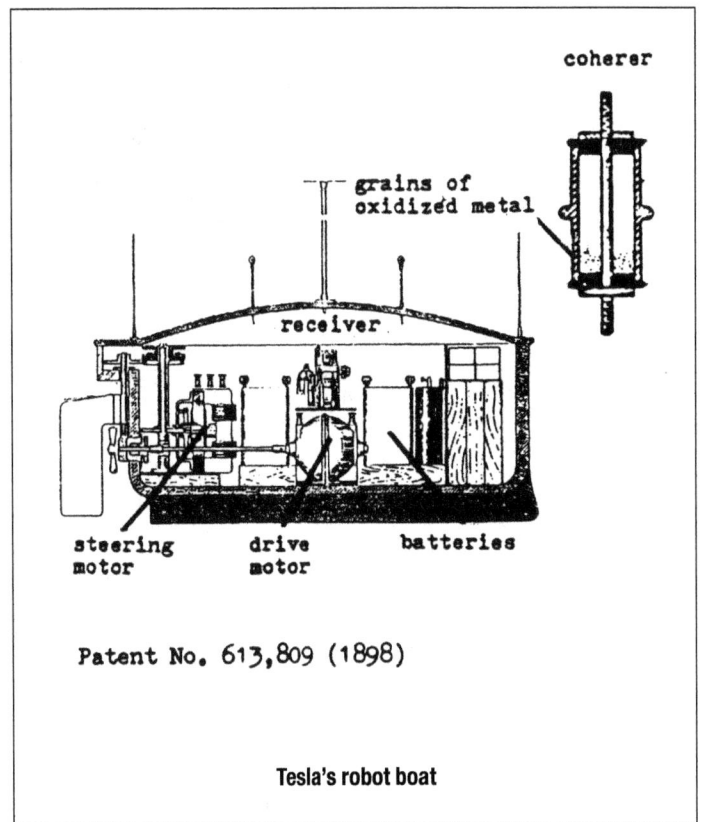

Patent No. 613,809 (1898)

Tesla's robot boat

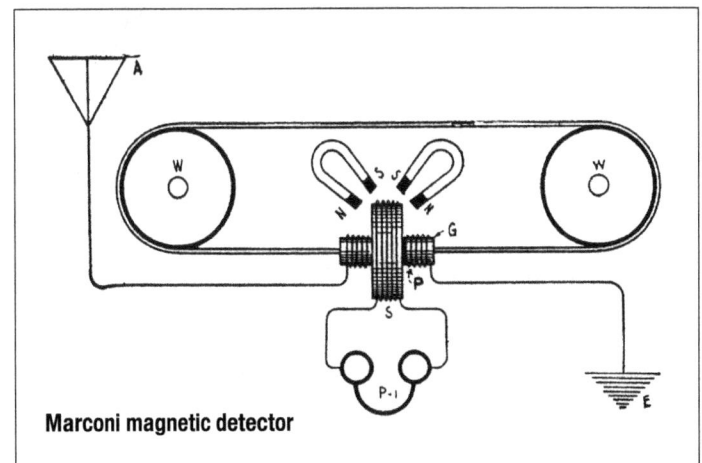

Marconi magnetic detector

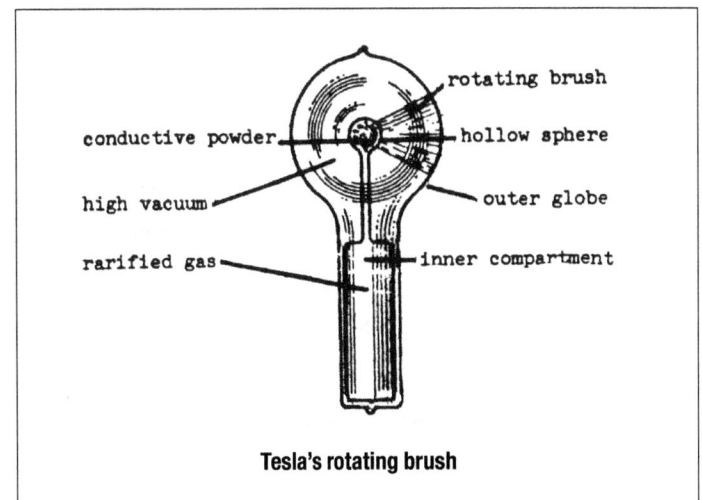

Tesla's rotating brush

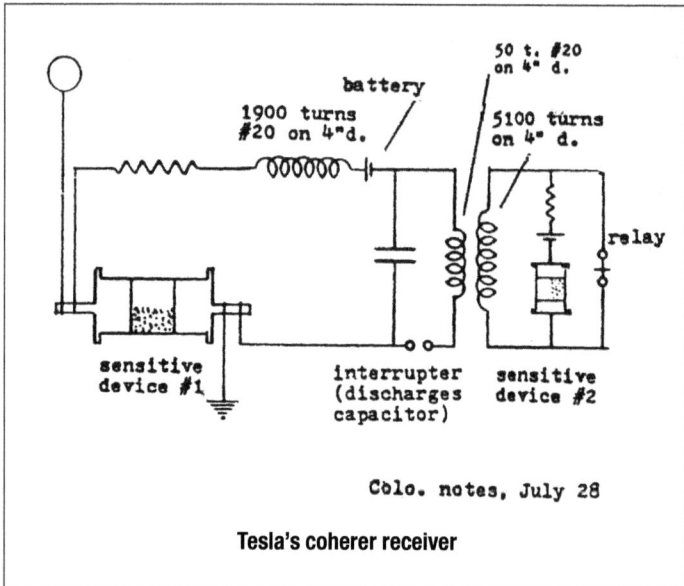

Tesla's coherer receiver

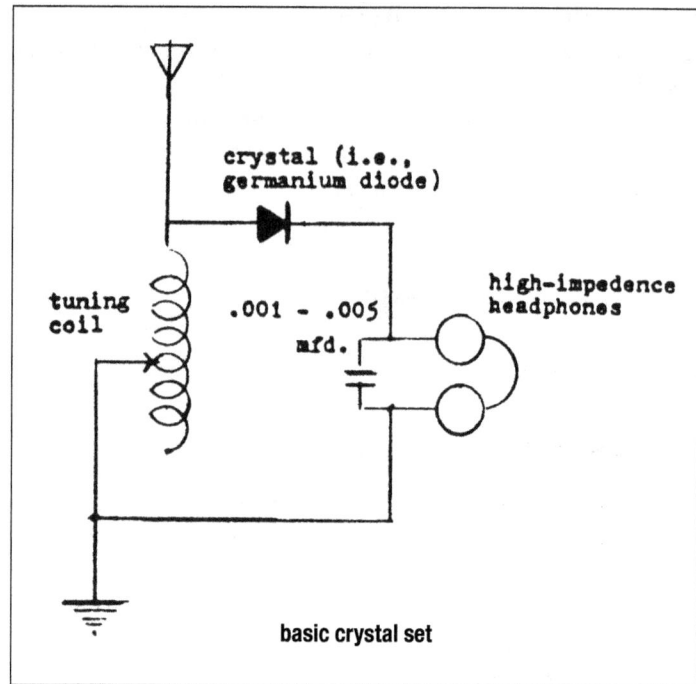

basic crystal set

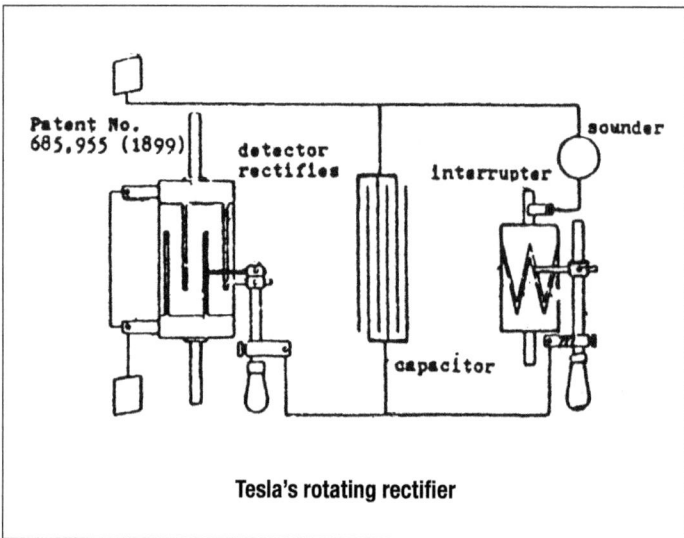

Tesla's rotating rectifier

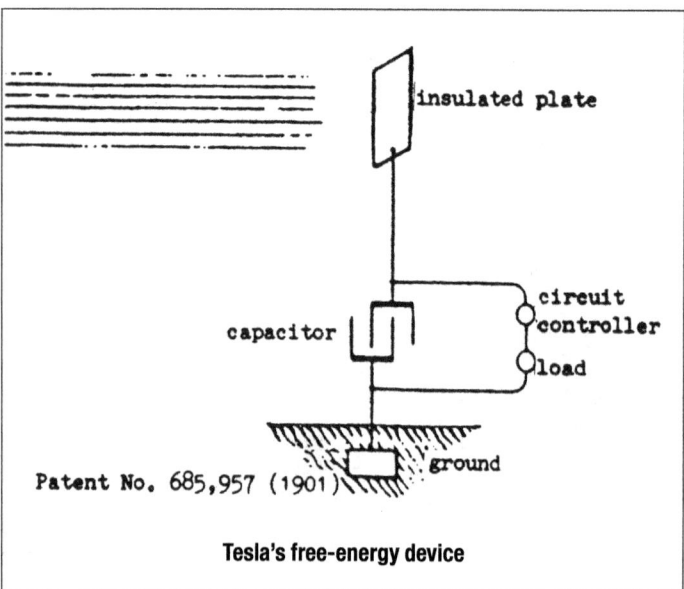

Tesla's free-energy device

Most of Tesla's receivers have one or more continuously moving parts, an offensive idea in this age of solid state. Tesla's coherer rotated, and so did his rectifying diode. Central to Tesla's receiver designs is another continuously moving part, a rotary "break" or switch used to discharge the capacitor at the proper intervals. The break discharges the capacitor through a sounder. It is a device that magnifies effects.

The moving parts notwithstanding, Tesla allows for semiconductors to do the job of the break. "The devices," he says in an 1898 patent, "may consist of merely two stationary electrodes separated by a feeble dielectric layer of minute thickness." Tesla's experience with spark gaps must have attuned him to the possibilities. He writes of thin insulative films serving as dielectrics. The patents offer no drawings or verbal detail. How far Tesla took these thin-film semiconductor ideas experimentally I do not know.

radio-free energy

Central to Tesla's receivers was the use of capacitors to store and release energy and to magnify effects: "However feeble or attenuated the impulses received, enough energy may be accumulated from them by storing up the energy of succeeding impulses for a sufficient interval of time to render the sudden liberation of it highly effective in operating a receiver." By "receiver" here Tesla means sounder. It probably took a fair amount of energy to drive the sounder, especially if what Tesla had in mind was the old magnetic click sounder used in wired telegraphy.

Tesla's free-energy patent (685,957) was filed at about the same time (1901) as a string of his radio receiver patents and is, in fact a kind of radio receiver. As in his radio receiving circuits Tesla is using precisely tuned capacitive discharges to magnify effects. Timing is tuning and is critical whether receiving signals or collecting energy at levels sufficient to do work.

The capacitors suitable for energy applications, Tesla says, should be "of considerable electrostatic capacity," and the dielectrics made of "the best quality mica."

the crystal set

The simplest detector diode is the crystal. It is a free-energy sensitive device with a long and illustrious history in radio, but there is no record that Tesla used it. It is not clear how close Tesla came to this kind of solid-state rectifying diode. He may have gone from the rotating rectifier to the vacuum diode unaware of the crystal principle, but I doubt it.

The crystal set is the simplest receiver in radio, and the tradition of crystal-set building persists even into this "high tech" era wedded to impressive complexity.

You can build a crystal set using an inexpensive germanium diode. The traditional rock crystal with cat's whisker is still available from Antique Radio Supply, as is the high-impedance headset that enables you to listen to crystal and other elemental receivers without amplification. The cat's-whisker crystal was mass-produced in the Crystal Age enclosed in a glass envelope and with the cat's whisker welded to the active spot. Although germanium became the dominant rectifying material for diodes, similar detectors show up in the literature where the same rectification is accomplished with less exotic materials: an acid solution, an electrolytic solution, a piece of strap iron. Soon after WWII there appeared in the literature a "foxhole" radio which used for its detector a safety pin and a razor blade of the no-longer-available blued-steel type. I built one, and it worked.

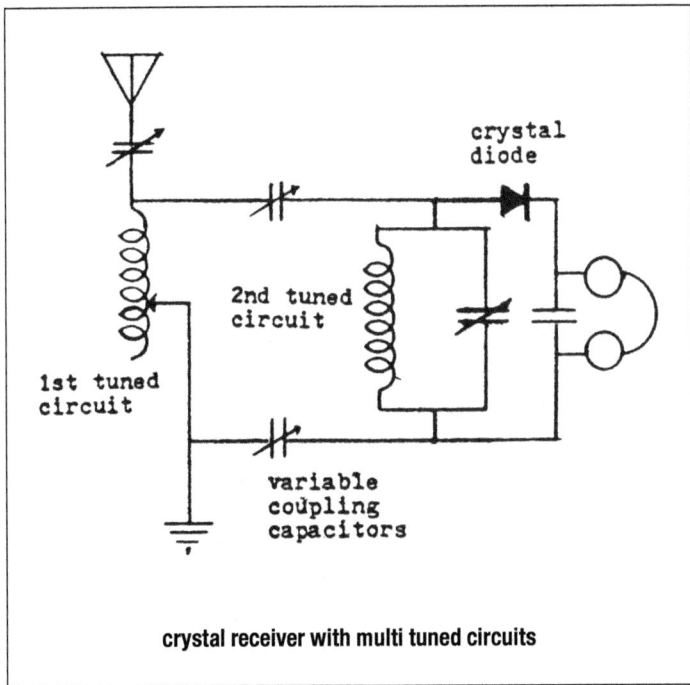

crystal receiver with multi tuned circuits

Some Diode Detectors

A. galena crystal and cat's whisker

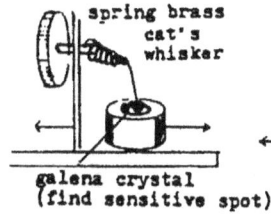

B. strap-iron detector

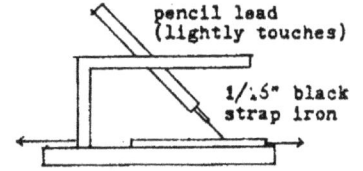

C. fox-hole radio

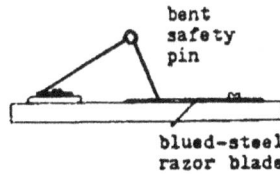

D. electrolytic detector

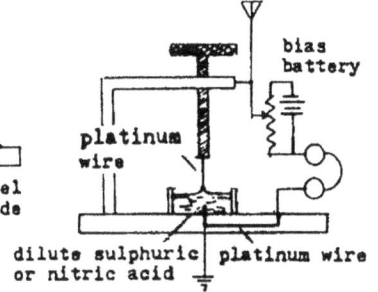

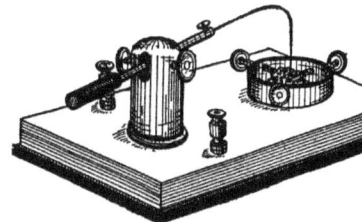

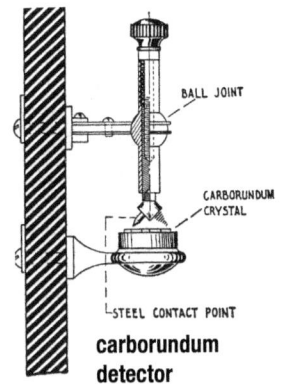

galena detector

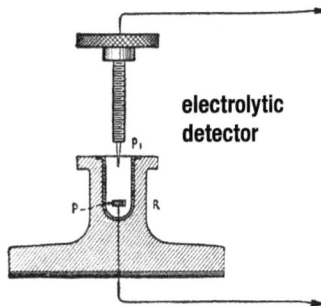

electrolytic detector

carborundum detector

zincite-bornite detector

germanium diode, 1N34

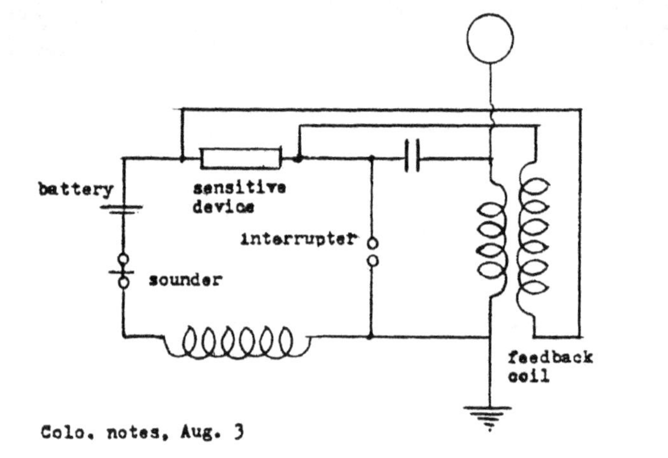

Colo. notes, Aug. 3

Tesla's regenerative receiver

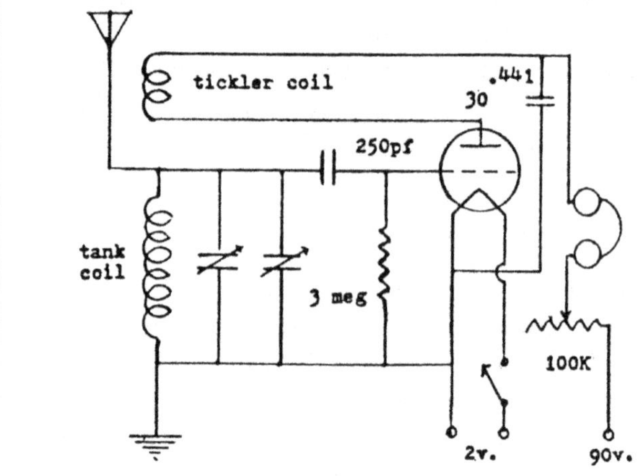

vacuum-tube regenerative

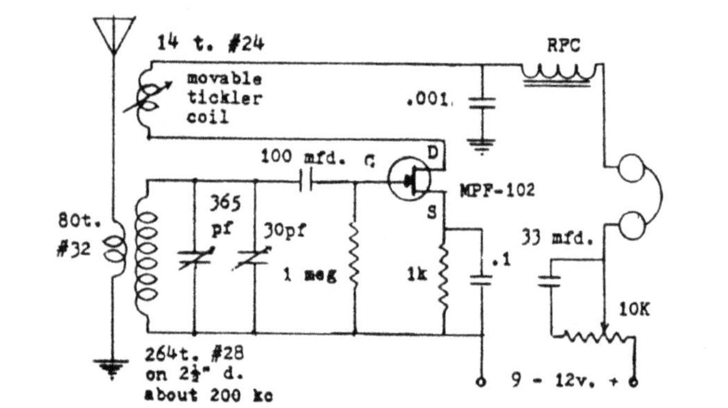

**author's low-frequency regenerative
how to build: see page 106.**

The crystal set is the simplest receiver you can build. While a single tuning (tank) circuit might work in the country, in the city one strong station may overwhelm all others unless multiple tuning circuits are used. A crystal set with two or more tuned circuits is the simplest practical radio receiver.

the regenerative

Tesla laid the foundation for the regenerative receiver. The vacuum-tube version, credited to Edwin Armstrong, was the receiver that succeeded the crystal set in the development of radio. "Regenerative" refers to the recirculating of the signal vibrating in the coil back into the sensitive device so as to further excite it. This feedback is another of Tesla's strategies for magnifying effects, and he knew about it early on from the self-excitation circuits he used in some of his dynamos.

The regenerative feedback link in Tesla's regenerative is a coil adjacent to the main tuning coil. In the vacuum-tube regenerative receiver, this became known as the "tickler coil." In some regen circuits the feedback link was through a capacitor, which could be a variable. Tesla did not patent the regen but shows many circuit variations of the idea in his Colorado notes, where he observes that it had "many valuable uses since by its means effects too feeble to be recorded in other ways may be rendered sufficiently strong to cause the operation of any suitable device."

The regen is one of the forgotten magical circuits of radio, but only recently forgotten. Familiar to every student of radio at least up through the '60s, it is not to be found in recent editions of *The Radio Amateurs Handbook*.

Although a regen often has a stage or two of amplification, an unamplified, single-stage regen having only a dozen parts is an effective listening tool for a wide range of frequencies from low through short wave. (As in other elemental receivers, high-impedance headphones are used. These differ from other phones in that there are many more turns of finer wire on the electromagnets, a technique to magnify effects, as in Tesla's magnetic detector.) The regen is sensitive and discriminating.

As is, without any additional circuitry, a regen can receive CW (code) and both AM and single-sideband phone.

In my youth I built a one-tube, 4-band, shortwave regen. Two years later I replaced it with a very basic commercial 4-band superhet (a National SW-54). The new National, to my dismay, could pull in little more than my little home-built regen. The only difference was in loudness and convenience of operation. The National had a calibrated dial and was easier to tune, for a regen is always a bit tricky, and the National had a rotary switch to select bands instead of plug-in coils. The National did not have an adjustable BFO, just a CW switch, so it did not match the regen in receiving the single-sideband phone signals that dominate the ham bands today.

My latest regen has one transistor. In Part 3, Tesla DIY you will find instructions for building this project.

Tesla's superhetrodyne

Tesla anticipated the superheterodyne receiver with his beat receptor. This is another of his electro-mechanical receivers. A taut steel band is set vibrating at "an enormous rate," says Tesla, by means of an electromagnet driven by a small reciprocating AC generator. This vibrating band is the local oscillator. Next to it is another electromagnet, small, sensitive, with many turns of fine wire. It functions as a pick-up for the steel band's vibrations, and through its winding also courses the received signal. The beating, or hetrodyning, makes the signal audible.

lament on solid state

In 1962, near the end of the era of above-ground nuclear testing, the U.S. government exploded a nuke of modest kilotonage above the atmosphere, that is, somewhere above 40 miles up. Several hundred miles away in Oahu, Hawaii, the electric pulse generated by the blast blew out street lights, fused powerlines, exploded TV sets and wrought other forms of electrical havoc.

CAUTION: STATIC-SENSITIVE DEVICE

label on Radio Shack IC package

No electrical device is totally safe from the impact of an electromagnetic pulse, or EMP, as it's called, which, if created by an above-atmosphere nuke, can load up power lines to 50 kilovolts per meter. An EMP can also be generated by lightning stroke, by a solar flare, by a Tesla magnifying transmitter, and by capacitive discharge, such as by a Marx impulse generator. In solid-state devices, the delicate dielectrics of integrated-circuit capacitors are particularly prone to vaporization, as are the junctions and tenuous conductors. Vacuum-tube devices are said to be ten-thousand times more resistant.

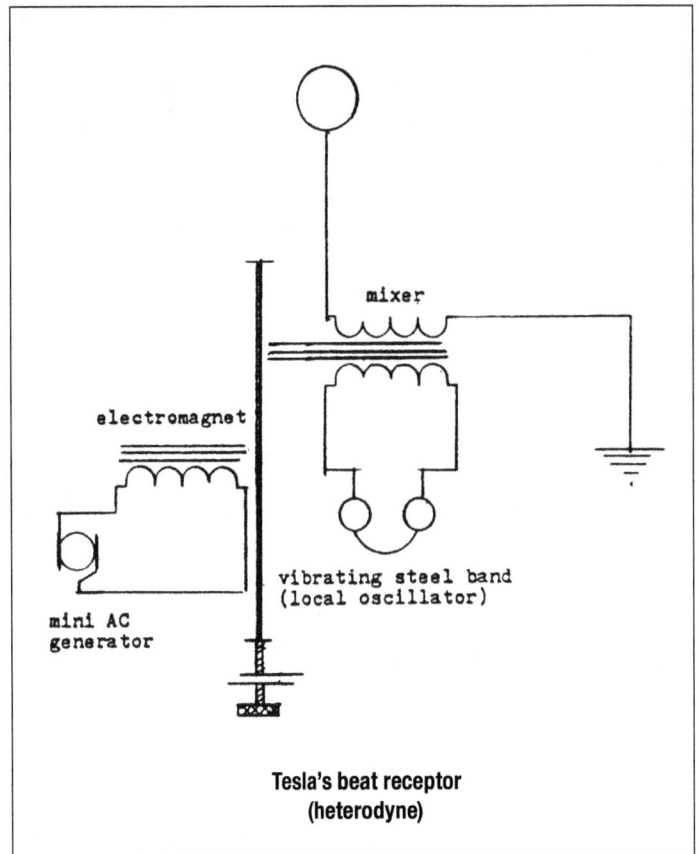

**Tesla's beat receptor
(heterodyne)**

It's been speculated that a nuke exploded 200 miles above Nebraska would dud all unprotected solid state circuitry in the continental United States. Particularly vulnerable are components connected to the power grid, to telephone lines, and to antennas. Magnetic memory could also be erased. While large institutions have been cued to this contingency and are moving ahead with the hardening of computer and communications facilities, the general public is largely oblivious to the fact that the entire high-tech electronic culture is EMP-destructible at a stroke.

This same vulnerability of solid-state devices to shock makes them a headache for the builder-experimenter. The ability to plug these little items into bread-boards is a big step forward in convenience, but, when a circuit fails, the experimenter is left wondering whether his hook-up is flawed or has the IC or transistor blown out due to excess heat in soldering, some minuscule excess of current or voltage, from reversed polarity, some mini-EMP-like static discharge, or kickback from some related high-voltage component.

The same cheap mass production of transistors and IC's that has made possible the world of digital has also encouraged the corruption of the quality of electronic components across the board. Switches, pots, audio transformers, variable capacitors, once built with integrity, have become cheaply made mini-junk, and this is often the only stuff readily available to experimenters. Parts suitable for building transmitters, Tesla coils, and other high-voltage, high-current work — like power transformers, heavy-duty wire-wound pots, chokes, high-watt resistors, transmitter variables, vacuum tubes, and insulators – must be obtained from surplus sources.

For experimenters mini-junk electronics also means tiny, brittle, vexatious, finger-puncturing, eye-straining connection terminals where there used to be hefty lead-in wires or sturdy posts. Miniaturization did not become an engineering obsession until the 1950's and the advent of the miniature and "acorn" tubes. The fashion ultimately reduced the size of all components to the minimum, including coils. But formerly, the size of a coil was correlated with its power. "Note the difference in size," said an ad in *The Electrical Experimenter* of 1917. The ad illustrated a 15,000-meter ham antenna-loading coil against a competitor's smaller counterpart. The advertised coil diameter was 10 inches, its length 32. A typical Tesla receiving coil was 25 inches in diameter. Tesla built with rugged components and on a shameless scale.

CHAPTER 6

Aerial Capacity

That ball sticking up in the air that is so symbolic of Tesla's radio: What is it? Some sort of antenna? Actually, Tesla never referred to it as an "antenna" but as an "aerial" or "air capacity" or as an "elevated capacity." Tesla did not see the elevated ball as a radiator, which is how the transmitting antenna of conventional Hertzian radio is construed. The aerial capacity corresponds to the terminal capacitor of a Tesla coil. Tesla said that in radio the aerial capacity "heightened the effect" of what is essentially a grounded system. Especially in transmitting, it appears that the aerial capacity provides a capacitive leverage against which to pump ground.

The ball shape also holds high voltage, minimizing coronal discharge and loss, and the ball aerial capacity at Colorado Springs was insulated to a high dielectric strength by a thick coating of rubber. The ball aerial appears on Tesla's receivers as well as his transmitters. Tesla understood that a long wire (as in long-wire antenna) had capacity, but he believed the sphere was a more efficient geometry. "By using a body of considerable surface better results are obtained than a wire leading to a height alone. The system is more economical in providing an electrical vibration in the ground." This would be especially true at low frequencies that would require a wire of inordinate length.

Tesla's Colorado transmitter when operating at 60 kc would have required a half-wave wire antenna 2500 meters long, if modern antenna conventions had been observed. Tesla also believed that the idea of "polarizing," or putting into parallel, transmitting and receiving antennas was nonsense. On his Colorado magnifying transmitter, Tesla used a hollow copper sphere only 30 inches in diameter. The ball geometry reduces to a minimum the problem

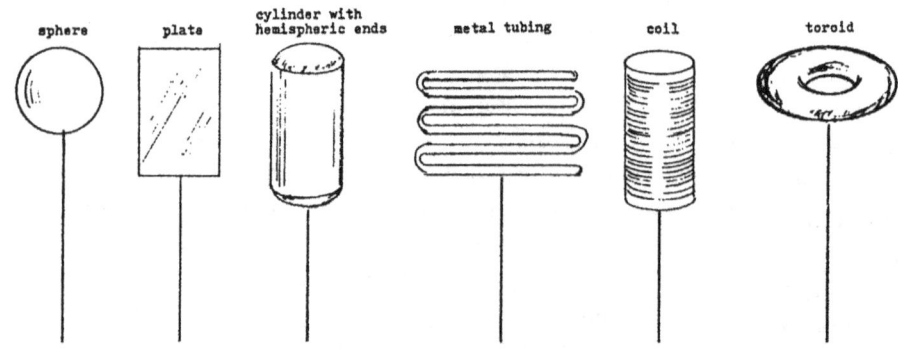

aerial capacity geometries

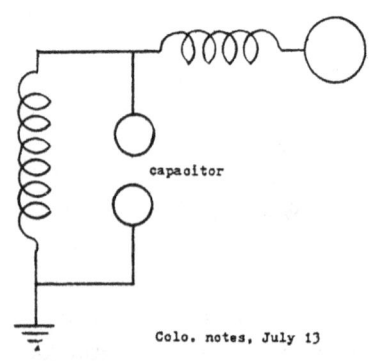

Tesla's opposing-ball capacitor

of streamers breaking out at high electrical pressures, since those jump more readily from angular surfaces. Tesla experimented extensively with the effect of the height of the ball. An increase in height caused an increase in the effective capacity of the ball, Tesla discovered. (However, of Hertzian radio, Tesla said, "The actions at a distance cannot be proportionate to the height of the antenna or the current in same.")

The ball worked for Tesla but so did other geometries. In Colorado Tesla experimented with a structure of iron pipes as an aerial capacity. He said the aerial capacity could be a cylinder with hemispheric ends, or it could be a toroid. Tesla suggested that a coil of insulated wire put aloft would suffice. He said any hollow vessel, like a ball, could be filled with a gas like hydrogen at low pressure for better effect. Some of Tesla's receiver schematics show the aerial capacity as a simple metal plate.

The ball capacity shows up in other Tesla circuits besides aerial, consisting of a capacitor of two opposing hollow balls.

dielectric antennas

There is an insulative or dielectric side to the electric physics of any aerial capacity or antenna. This is why Tesla's spherical aerial capacity atop his Colorado tower was heavily coated with an insulation of rubber. Tesla would appreciate the world of plastics, ceramics, and all of the more exotic dielectrics available to today's inventor.

The dielectric properties of radio propagation go unappreciated in most discourse on radio. An exception is Eric Dollard, who speaks of dielectricity. There is an unexplored world in the dielectric side of all electric phenomena, including radio propagation. Indeed, there is a microwave item called the dielectric aerial.

A lesson in dielectrics came to me over-the-transom from a reader who had noted the free-energy receiver patent in Lost Inventions. He had found somewhere in the literature a diagram of a free-energy device analogous to Tesla's that he wanted to share. The schematic specified that the entire system, aerial to ground, be insulated to at least six-thousand volts. The capacitor dielectric was to be 6 KV or better.

I had been experimenting with the ambient-energy collecting power of various capacitors but had formed a bias toward the high-capacity electrolytic filter type, for I had happened upon an electrolytic ten inches tall rated a thirsty 10,000 mfd. and 450 volts (an almost unheard of voltage rating for an electrolytic). I assumed that this would be the performer. But I took the reader's suggestion and tried an even more exotic pulse capacitor rated only 2 mfd. but with a dielectric strength of two-thousand volts. Four in a series = 8 KV. It was considerably more effective at energy collection.

For Tesla the medium of radio propagation is an ether which he perceived to be at high tension. (Six KV?) The interface between antenna and ether can be seen as capacitive, dielectric.

It's logical that the most effective interface with this medium would be via a material of high dielectric strength.

This insight led to the discovery that, if you insulate an antenna to 6 KV or better, you can dramatically boost effectiveness in transmitting. My experiment used a field-strength meter. I compared the performance of a CB antenna made of 600-volt, #14 house wire with another made of 15 KV high-voltage-test wire (#11, stranded). To my surprise, the signal strength from the 6 KV wire registered 150 percent higher than the 600-volt wire.

CB'ers and hams commonly favor enamel or bare wire for aerials or naked aluminum tubing. Are they missing out? Meanwhile the dielectric advantage might explain why rubber-ducky aerials on hand-held transceivers work so well.

Whenever possible, I use high-voltage test wire for any antenna, for transmitting or receiving, and I would coat an aluminum antenna of any configuration with a dielectric layer by dipping or spraying.

capacitive antennas

The "capacitive hat" appears in modem antenna design. Although the texts explain it as something that "improves radiating efficiency," it is still in the nomenclature as a "capacitive hat." Capacitive hats appear frequently in low-frequency radio, in the antenna systems of navigational beacons, of LORAN, of GWEN, and among LowFERs. The capacitive hat appears on top of a helical antenna (which, as a coil of wire aloft, might qualify as an aerial capacity anyway). The loading coil on a top-loading whip antenna may be "seen" as an aerial capacity?

In the radio literature of the 1920s there are references to a "capacitive antenna." The capacity antenna is recommended as an indoor alternative for situations where an outdoor antenna is impossible. Even where an outdoor is possible, one may want to be more subtle. Conventional Hertzian antennas, like the ham array and the CB whip, are flags announcing the presence of a transmitter.

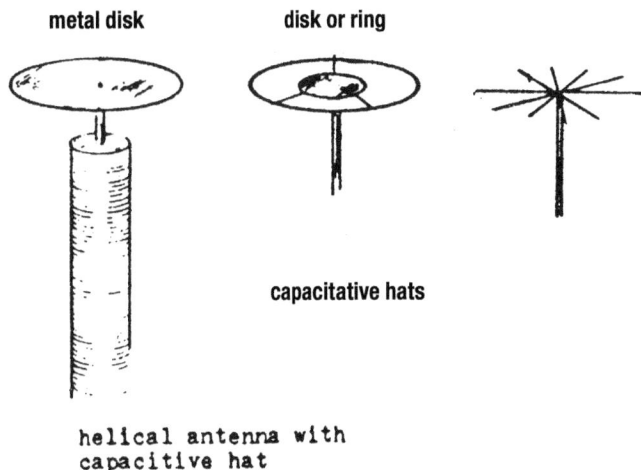

metal disk

disk or ring

capacitative hats

helical antenna with
capacitive hat

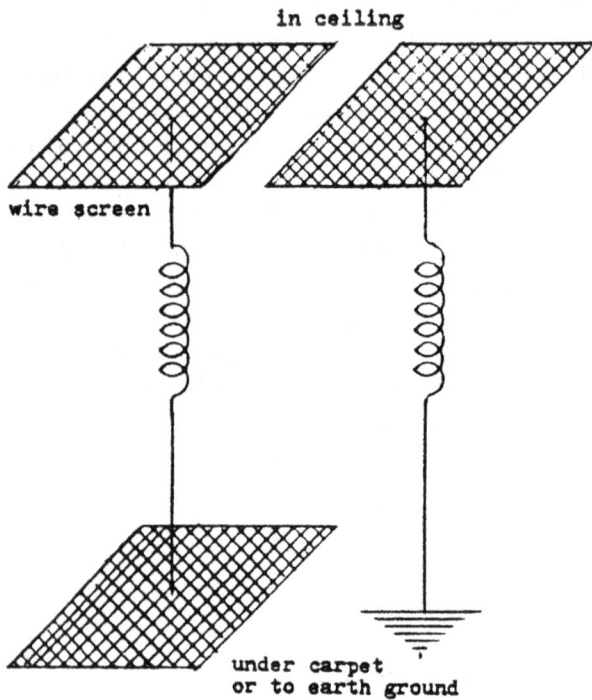

capacity antenna

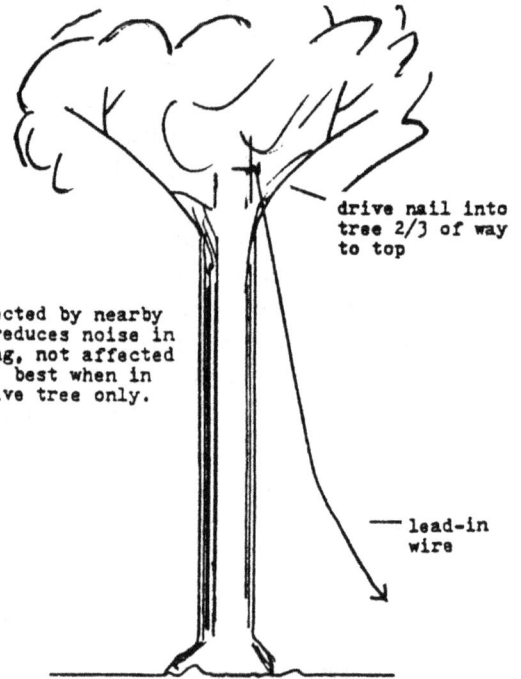

tree as antenna

in ceiling

wire screen

under carpet
or to earth ground

not affected by nearby
trees, reduces noise in
receiving, not affected
by rain, best when in
leaf, live tree only.

drive nail into
tree 2/3 of way
to top

lead-in
wire

The capacitive antenna could consist of a sheet of metal or a wire screen set aloft. Where a ground connection was distant, another screen was set below, the two separated by 10 to 15 feet. The upper element could be attached to the ceiling or placed in an attic, the lower under the carpet.

Loomis sent kites aloft as aerials; they were covered with copper gauze. Loomis also observed that a tree could be used as a receiving antenna, and others have discovered the tree as a transmitting antenna as well. I've seen no mention of frequencies in the tree-as-antenna literature, which comes from the low-frequency era. (I've tried a tree for receiving, and it worked on longwave, medium-wave, and shortwave bands.) Does the transmitter or receiver "see" the tree as a Tesla-style aerial capacity?

loops

Loop antennas abound in radio and suggest Tesla's elevated-coil idea; You can easily build your own using PVC tubing for the structure. For strength, use one-inch schedule 40 tubing instead of the 1/2 inch that I used (photo).

The shielded loop appears in the LowFER receiver literature as a noise-reduction strategy. The shielding excludes the magnetic component of disturbances while the loop inside responds only to the electrostatic, according to the literature. A preamp is a good idea with any loop.

studded mushroom

Tesla's magnifying transmitter patent shows a toroid-shaped aerial capacity. It is studded with half-spherical metal plates, presumably to enlarge the surface without increasing by much the bulk. The ultimate Tesla aerial capacity is the studded mushroom that topped his Wardencliff tower. Its diameter was 68 feet. The studding may have been parabolic rather than strictly hemispherical. There is not much information about the tower, and Tesla's reasoning behind the strange geometry is not fully understood.

The notion of aerial capacity has little currency outside of Tesla-land. What would antennas be like if they were reinvented, not as senders and receivers of Hertzian radiation, but as aerial capacitors?

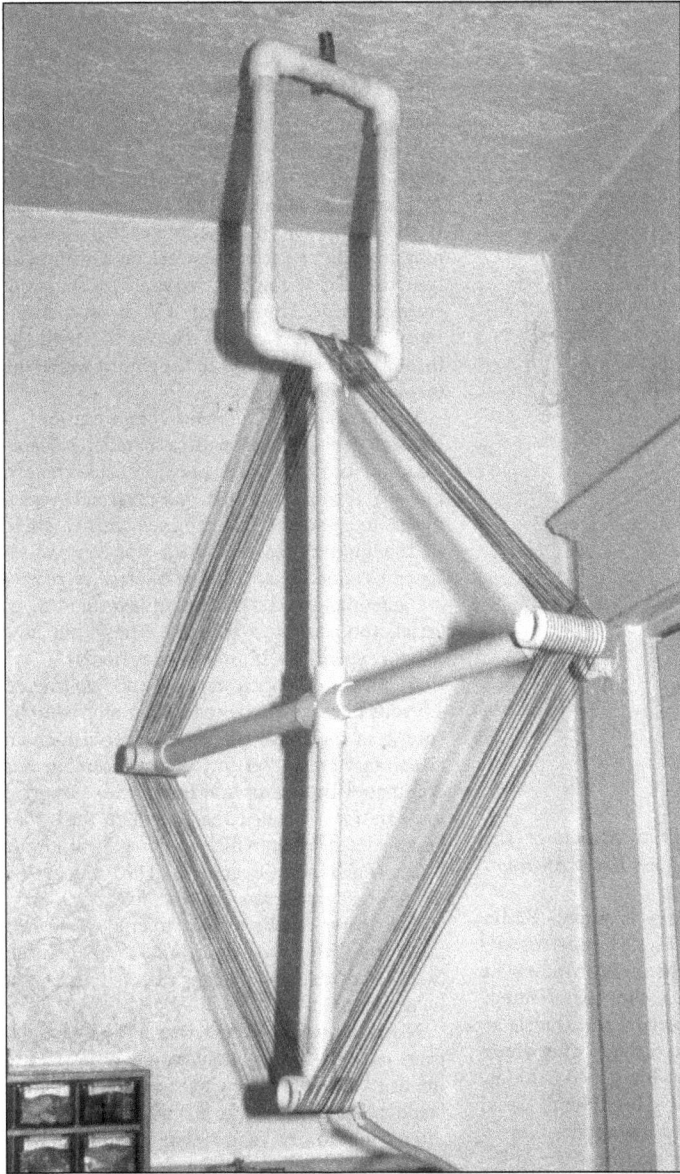

author's loop antenna

parabolic
metal
plates

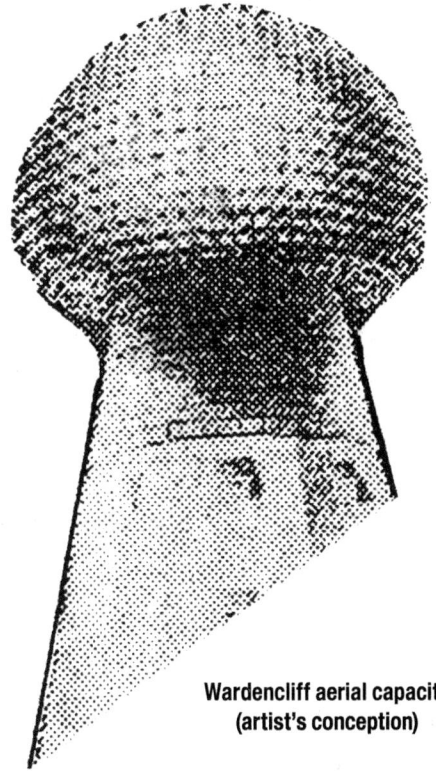

Wardencliff aerial capacity
(artist's conception)

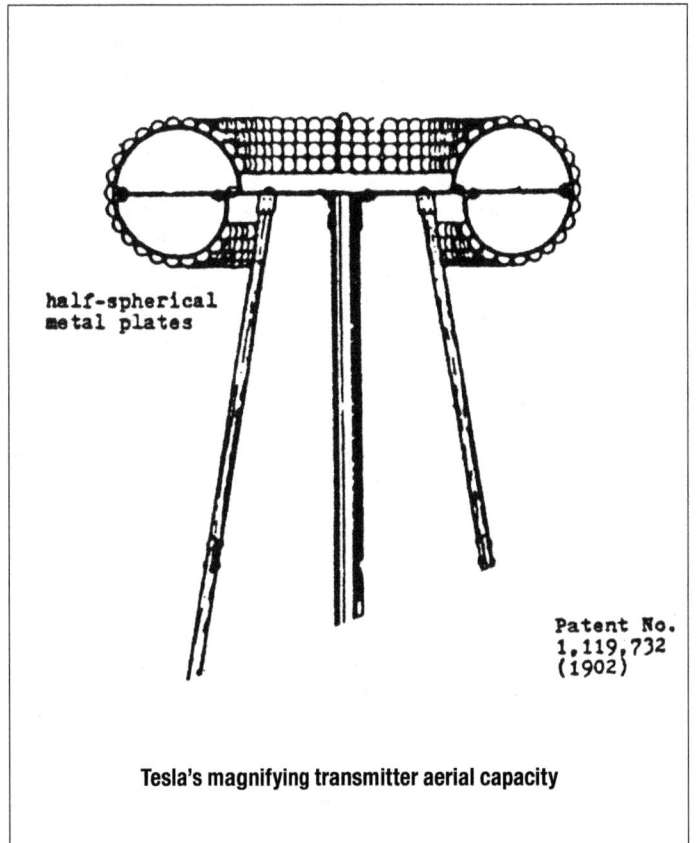

coax: cable tv
hardline, RG-59, etc.
Loop can also be many
turns in metal tubing.
Not necessarily circular.

open shield
1" at top

rotate loop for
minimum noise

length less than
.08 wavelength

metal
box

shielded loop

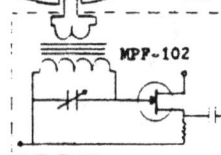

MPF-102

coax to
receiver

by Mark Mallory
See also QST 3/74

half-spherical
metal plates

Patent No.
1,119,732
(1902)

Tesla's magnifying transmitter aerial capacity

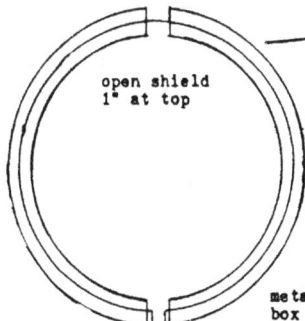

Appendix B

*everything you know
is wrong*

The True Wireless

by Nikola Tesla
edited by George Trinkaus

Introduction
by George Trinkaus

Tesla wrote *The True Wireless* back in 1919 because he wanted to knock the prevailing orthodoxy of radio theory into a cocked hat. I am republishing the piece in 2014 because that orthodoxy still prevails and because I agree with Tesla that into a cocked hat is where this persistent and annoying body of dogma should be knocked.

How does radio really work? What responsible scientific explanations can we propose for the mysteries of wireless? What are the true dynamics of propagation in as much as we can ever know them?

Unfortunately, early on in the development of radio, explanations were proposed, and, as is the sad habit of academic science, some of these conjectures readily solidified into the official theory.

Does modem science have no patience with mystery? In 1901 long-distance wireless was achieved. In the very next year, a respected scientist, Oliver Heaviside, suggested this might be done by signals bouncing off of ionized layers presumed to exist a hundred or so miles up. Perhaps because it was consistent with the already accepted notion that radio behaved like light rays that can be reflected, the Heaviside layer (later called the ionosphere) had instant academic appeal, and, despite the protest of thinkers with other ideas about how radio worked, the theory of ionospheric propagation became a fixture in the official theory, enabling radiomen through the century to speak glibly of "skip," whether any such a phenomenon demonstrably exists or not. Tesla said, "There is no Heaviside layer, or, if it exists, it is of no effect."

Here in a nutshell is the official theory of wireless propagation in the ham bands as abstracted from the 1996 edition of the American Radio Relay League's *Radio Amateur's Handbook*:

Radio is radiation. Radio is waves. Radio waves behave like light rays. Unobstructed, as in the "vacuum" of space, radio moves in straight lines. Like light it can be refracted, reflected, scattered, or absorbed and travels at exactly 186,000 miles per second. The whole model of orthodox radio is demystified through this analogy to the observable behavior of light.

The layers of the so-called ionosphere vary in density, and, hence refractivity. Under the influence of sunlight, there are day-night fluctuations and 11-year sun-spot cycles which enhance or diminish ionospheric reflection and hence the ability of short-wave radio to propagate long-distance in this fashion.

Radio is assumed to be an aerial phenomenon. Although a surface-traveling "ground wave" may be acknowledged, conduction through the earth, which Tesla proposes, is assumed to be an impossibility in the orthodoxy. To Tesla, radio propagation is conduction.

Tesla says here that orthodox Hertz radio theory is "one of the most remarkable and inexplicable aberrations of the scientific mind which has ever been recorded in history," and that it "has stifled creative effort in the wireless art and retarded it for twenty-five years."

Orthodoxy drives out speculation. We can be grateful to Tesla for his attempt to smash this orthodoxy and for his giving us, if we go with him, a new beginning in this neglected science. Liberated, we can proceed to wonder again at the mysteries, to speculate, experiment, even have some fun.

an experiment

Mostly for the last reason, I built a sensitive field-strength meter, connected it between an antenna and ground, tuned it to a strong local AM broadcast signal, and sat back and watched. I used a large-face microammeter that could be read from fifteen feet away. Set on a worktable near the foot of my bed, it became a source of entertainment, my etheric TV. It was essentially the simplest crystal receiver but with the 0-to-200 microammeter in the circuit where the earphones normally go.

For Tesla-style resonant amplification, I connected a big coil with a parallel variable capacitor as a tank tuner between field-strength meter and ground. I had also tried it between meter and antenna but achieved greater effect in the ground circuit, which was suggestive. (If you connect your AM or shortwave receiver's antenna terminal to ground instead of to an aerial, you commit a Hertzian heresy, but how do you explain the improved reception?)

As measured on my vacuum-tube voltmeter, 1.5 volts or more can be coaxed out of ground by tuning in the above fashion. Typical of such an elementary crystal set in the city, reception was dominated by one or another of two powerful local stations. I tuned the system to the strongest meter reading, which gave me 50-kilowatt KEX (Portland, Oregon) at 1190 kilocycles, whose transmitter was less than five miles away.

The meter needle swung to mid-range and held, twitching to the signal's modulation (which, surprisingly, was less than ten percent of its deflection).

Now one would expect that a 50 kw signal from only five miles away would be of steady intensity, but variations over a period of days could be as dramatic as 50 microamps. It was interesting to try to correlate variations with

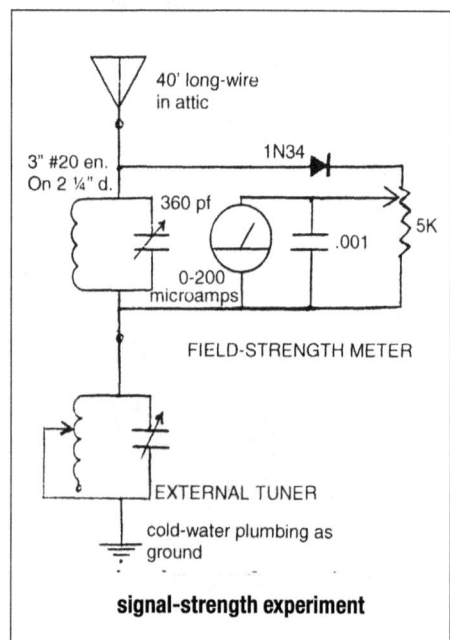

signal-strength experiment

weather conditions, day and night, lunar cycles, etc., but none of these speculations was conclusive. What was conclusive was that nothing in conventional propagation theory accounted for what I was seeing.

Hertzian theory does allow for propagation at this relatively low frequency by means of the so-called ground wave. Thus any variations in the alleged ionosphere would not be a factor, nor would it be anyway at this short distance.

invitation to speculation

So here is an instance of how the orthodoxy is empty. And here is an invitation to speculate anew. What is the true nature of the radio-conductive medium? I don't know. You don't know. We can only guess. Evidently the medium is rich in influences. The atmosphere and the earth (let's call it all the electrosphere) may be coursed by currents, not all electrical necessarily but also of other "unusual energies," including perhaps those known to the dowser. These energies might exercise attenuating and even enhancing influences on the radio energies imposed by man.

So let's hypothesize that the medium could be a veritable soup of churning energetic activity that can affect a signal. Let's appreciate our ignorance, be humble, and refrain from smug assertions, dear ARRL and its graduates.

unpublishable

By June 1919 when Hugo Gemsbach published "The True Wireless" in his magazine, *The Electrical Experimenter*, Tesla was well on his way to being rendered invisible by the media, and he must have been grateful for this opportunity to express himself publicly.

In earlier years, the media had romanced Tesla to the public as the "electrical genius" who had invented the alternating-current system, and Tesla had enjoyed great access to the press. But when Tesla reinvented his own system, after it had been established industrially as the official system, his voice had to be silenced.

By 1921 Tesla was almost unpublishable, officially at least. In the few years prior to that, *The Electrical Experimenter* had published Tesla profusely. Tesla's publication of "The True Wireless" in *The Experimenter* would appear to have been granted grudgingly, though, considering the way in which the article was crammed into just four pages. Because of this tight economy, and because of the limitations of metal-type layout in the letterpress era, the illustrations in the original are often placed remotely from the relevant text. Moreover, many of the original illustrations are tiny. All this makes the reading of the original a drag. I trust that my drudgeries in resetting and redesigning this work make your experience a more pleasant one.

In this writing, Tesla, aware that in respect to the establishment he had nothing to lose, indulges a candor that he could not previously "on the ground," he says, "that such idle and far-fetched speculations would injure me in the opinion of conservative businessmen."

energy projection

Tesla's far-fetched, speculative wireless is based on an electric physics that allows for the phenomenon of projected electric energy. Tesla is using electrician parlance when he calls this phenomenon "single wire without return." Tesla is determined that this radical novelty be comprehended by his brain-washed readers and he resorts to illustrations using crude hydraulic

analogs (Figures 4-6). If you have seen a hot Tesla coil in action, you have been shown a more poetic manifestation of Tesla's electrics in the corona, sparks, and streamers that leap from the terminal, eagerly seeking conduction in the ambient medium. The single-wire-without-return phenomenon is at the heart of Tesla's radio and wireless power.

Done Tesla's way, the power of radio is "unlimited." But, says Tesla, "the experts are blind to the possibilities." This article "is Tesla's lament on this theme. The tone suggests the frustration of the inventor who has been standing by mutely over the years seeing his radio used but at the same time misused and the phenomenon systematically misexplained even to its practitioners.

No ionosphere. Not an aerial phenomenon. Antenna current irrelevant. Parallel antennas unnecessary. Mountains irrelevant. Short waves ineffective. What heresies! Where would the average ham be without the precepts Tesla challenges here? The ham license exam requires at least a rote grasp of the orthodoxy. The ham's social milieu takes the orthodoxy for granted.

But the old dogma does not serve us, says Tesla. It is the theory of an impotent radio. "The transmitter generates several systems of waves all of which, except one, are useless," he has said elsewhere, suggesting that a conventional transmitter works only inadvertently. The result is what he has called an "unfit apparatus" for the job.

Tesla's boasts that his radio is so potent that "transmission is absolutely unlimited as to terrestrial distance and the amount of energy," which means energy sufficient not only for communications but for the lighting of cities and the driving of the machinery of industry.

This, says Tesla, is the true wireless.

The True Wireless

by Nikola Tesla (1919)

Ever since the announcement of Maxwell's electromagnetic theory, scientific investigators all the world over had been bent on its experimental verification. They were convinced it could be done and lived in an atmosphere of eager expectancy, unusually favorable to the reception of any evidence to this end. No wonder then that the publication of Dr. Heinrich Hertz' results caused a thrill as had scarcely ever been experienced before.

At that time I was in the middle of pressing work in connection with the commercial introduction of my system of power transmission, but, nevertheless, caught the fire of the enthusiasm and fairly burned with desire to behold the miracle with my own eyes. Accordingly, as soon

as I had freed my self of these imperative duties and resumed research work in my laboratory on Grand Street, New York, I began, parallel with high-frequency alternators, the construction of several forms of apparatus with the object of exploring the field opened up by Dr. Hertz.

oscillation transformer

Recognizing the limitations of the devices he had employed, I concentrated my attention on the production of a powerful induction coil but made no notable progress until a happy inspiration led to the invention of the oscillation transformer. In the latter part of 1891 I was already so far advanced in the development of this new principle that I had at my disposal means vastly superior to those of the German physicist. All my previous efforts with Rhumkorf coils had left me unconvinced, and in order to settle my doubts I went over the whole

ground once more very carefully, with these improved appliances. Similar phenomena were noted, greatly magnified in intensity, but they were susceptible of a different and more plausible explanation.

Tesla visits Hertz

I considered this so important that in 1892 I went to Bonn, Germany, to confer with Dr. Hertz in regard to my observations. He seemed disappointed to such a degree that I regretted my trip and parted from him sorrowfully. During the succeeding years I made numerous experiments with the same object, but the results were uniformly negative.

In 1900, however, after I had evolved a wireless transmitter which enabled me to obtain electromagnetic activities of many millions of horsepower, I made a last desperate attempt to prove that the disturbances emanating from the oscillator were ether vibrations akin to light, but met again with utter failure.

For more than eighteen years I have been reading treatises, reports of scientific transactions, and articles on Hertz wave telegraphy, to keep myself informed, but they have always impressed me like works of fiction.

perishable theories

The history of science shows that theories are perishable. With every new truth that is revealed we get a better understanding of Nature, and our conceptions and views are modified. Dr. Hertz did not discover a new principle. He merely gave material support to a hypothesis which had been long ago. formulated.

It was a perfectly well established fact that a circuit, traversed by a periodic current, emitted some kind of space waves, but we were in ignorance as to their character, He apparently gave an experimental proof that they were transversal vibrations of the ether. Most people look upon this as his great accomplishment. To my mind it seems that his immortal merit was not so much in this as in the focusing of the investigator's attention on the processes taking place in the ambient medium.

The Hertz wave theory, by its fascinating hold on the imagination, has stifled creative effort in the wireless art and retarded it for twenty-five years. But, on the other hand, it is impossible to over-estimate the beneficial effects of the powerful stimulus it has given in many directions.

As regards signaling without wires, the applications of these radiations for the purpose was quite obvious. When Dr. Hertz was asked whether such a system would be of practical value, he did not think so, and he was correct in his forecast. The best that might have been expected was a method of communication similar to the heliographic and subject to the same or even greater limitations.

single wire without return

In the spring of 1891 I gave my demonstrations with a high-frequency machine before the American Institute of Electrical Engineers at Columbia College, which laid the foundation to a new and far more promising departure. Although the laws of electrical resonance were well known at that time (and my lamented friend Dr. John Hopkinson had even indicated their specific application to an alternator in the Proceedings of the Institute of Electrical Engineers, London, Nov. 13, 1889), nothing had been done towards the practical use of this knowledge, and it is probable that those experiments of mine were the first public exhibition with resonant circuits, more particularly of high frequency.

While the spontaneous success of my lecture was due to spectacular features, its chief import was in showing that all kinds of devices could be operated through a single wire without return. This was the initial step in the evolution of my wireless system. The idea presented itself to me that it might be possible, under observance of proper conditions of resonance, to transmit electric energy through the earth, thus dispensing with all artificial conductors.

conservative businessmen

Anyone who might wish to examine impartially the merit of that early suggestion must not view it in the light of present day science. I only need to say that as late as 1893, when I had prepared an elaborate chapter on my wireless system dwelling on its various instrumentalities and future prospects, Mr. Joseph Wetzler and other friends of mine emphatically protested against its publication on the ground that such idle and far-fetched speculations would injure me in the opinion of conservative businessmen.

So it came that only a small part of what I had intended to say was embodied in my address of that year before the Franklin Institute and the National Electric Light Association under the chapter "On Electrical Resonance." This little salvage from the wreck has earned me the title of "Father of the Wireless" from many well-disposed fellow workers, rather than my invention of scores of appliances which have brought wireless transmission within the reach of every young amateur and which, in a time not distant, will lead to undertakings over-shadowing in magnitude and importance all past achievements of the engineer.

The popular impression is that my wireless work was begun in 1893, but as a matter of fact, I spent the two preceding years in investigations, employing forms of apparatus, some of which were almost like those of today.

It was clear to me from the very start that the successful consummation could only be brought about by a number of radical improvements. Suitable high-frequency generators and electrical oscillators had to first to be produced. The energy of these had to be transformed in effective transmitters and collected at a distance in proper receivers. Such a system would be manifestly circumscribed in its usefulness if all extraneous interference were not prevented and exclusiveness secured. In time, however, I recognized that devices of this kind, in order to be most effective and efficient, should be designed with due regard to the physical properties of this planet and the electrical conditions obtaining in same.

I will briefly touch upon the salient advances as they were made in the gradual development of the system. The high-frequency alternator employed in my first demonstrations is illustrated in Figure 1. It comprised a field ring with 384 pole projections and a disk armature with coils wound in one single layer which were connected in various ways according to requirements. It was an excellent machine for experimental purposes, furnishing sinusoidal currents of from 10,000 to 20,000 cycles per second. The output was comparatively large, due to the fact that as much as 30 amperes per square millimeter could be passed through the coil without injury.

The diagram in Fig. 2 shows the circuit arrangements as used in my lecture. Resonant conditions were maintained by means of a condenser subdivided into small sections, the finer adjustments being effected by a movable iron core within an inductance coil. Loosely linked with the latter was a high tension secondary which was tuned to the primary.

The operation of devices through a single wire without return was puzzling at first because of its novelty, but can be readily explained by suitable analogs. For this purpose reference is made to Figures 3 and 4.

In the former the low resistance electric conductors are represented by pipes of large section, the oscillator by an oscillating piston, and the filament of an incandescent lamp by a

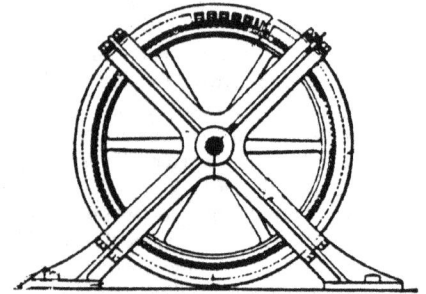

1. high-frequency alternator

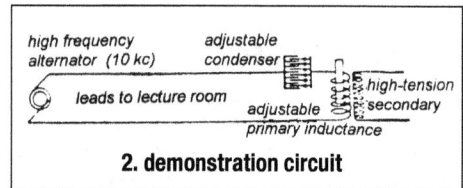

high frequency alternator (10 kc)

leads to lecture room

adjustable condenser

high-tension secondary

adjustable primary inductance

2. demonstration circuit

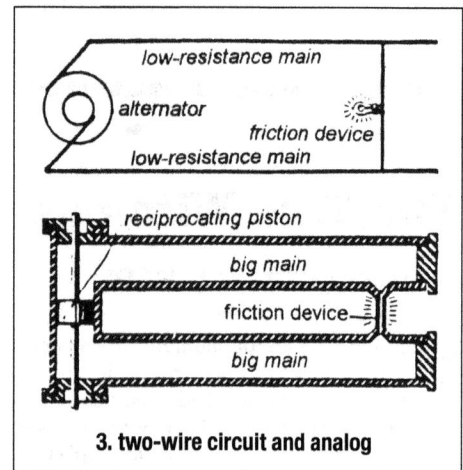

low-resistance main

alternator

friction device

low-resistance main

reciprocating piston

big main

friction device

big main

3. two-wire circuit and analog

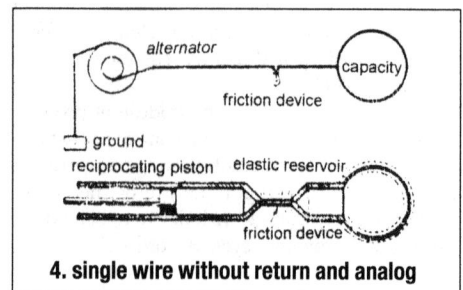

alternator

capacity

friction device

ground

reciprocating piston

elastic reservoir

friction device

4. single wire without return and analog

minute channel connecting the pipes. It will be clear from a glance at the diagram that very slight excursions of the piston would cause the fluid to rush with high velocity through the small channel and that virtually all the energy of movement would be transformed into heat by friction, similarly to that of an electric current in the lamp filament.

The second diagram will now be self-explanatory. Corresponding to the terminal capacity of the electric system, an elastic reservoir is employed which dispenses with the necessity of a return pipe. As the piston oscillates, the bag expands and contracts, and the fluid is made to surge through the restricted passage with great speed, this resulting in the generation of heat in the incandescent lamp. Theoretically considered, the efficiency of conversion of energy should be the same in both cases.

Granted, then that an economic system of power transmission through a single wire is practicable, the question arises how to collect the energy in the receivers. With this object attention is called to Fig. 5, in which a conductor is shown excited by an oscillator joined to it at one end.

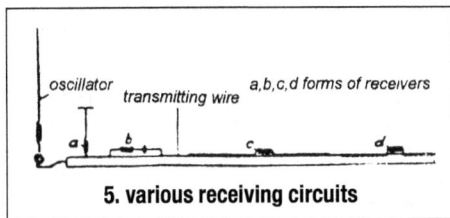

5. various receiving circuits

Evidently, as the periodic impulses pass through the wire, differences of potential will be created along the same as well as in right angles to it in the surrounding medium, and either of these may be usefully applied. Thus at a circuit comprising an inductance and a capacity is resonantly excited in the transverse, and at b in the longitudinal sense. At c energy is collected in a circuit parallel to the conductor but not in contact with it, and again at d, in a circuit which is partly sunk into the conductor and may be, or not, electrically connected to same. It is important to keep these typical dispositions in mind, for, however the distant actions of the oscillator might be modified through the immense extent of the globe, the principles involved are the same.

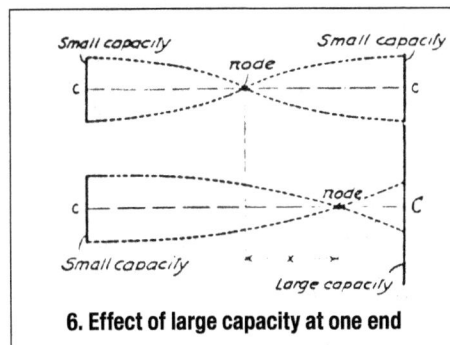

6. Effect of large capacity at one end

Consider now the effect of such a conductor of vast dimensions on a circuit exciting it. The upper diagram of Fig. 6 illustrates a familiar

oscillating system comprising a straight rod of self-inductance 2L with small center capacities cc and a node in the center. In the lower diagram of the figure a large capacity C is attached to the rod at one end with the result of shifting the node to the right, through a distance corresponding to self-inductance X. As both parts of the system on either side vibrate at the same rate, we have evidently (L + X), c = (L − X) C from which

$$X = L \frac{C - c}{C + c}$$

When the capacity C becomes commensurate to that of the earth, X approximates L, in other words, the node is close to the ground connection. The exact determination of its position is very important in the calculation of certain terrestrial electrical and geodetic data, and I have devised special means with this purpose in view.

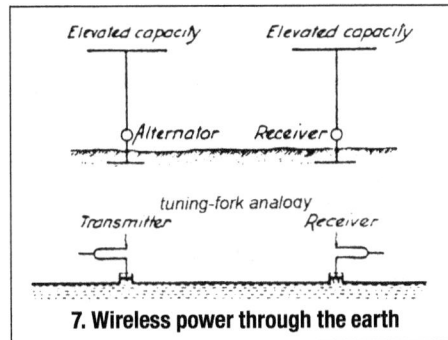

7. Wireless power through the earth

My original plan of transmitting energy without wires is shown in the upper diagram of Fig. 7, while the lower one illustrates its mechanical analog, first published in my article in the *Century* magazine of June 1900. An alternator, preferably of high tension, has one of its terminals connected to the ground and the other to an elevated capacity and impresses its oscillations upon the earth. At a distant point a receiving circuit, one of its terminals likewise connected to ground and the other to an elevated capacity, collects some of the energy and actuates a suitable device. I suggested a multiplication of such units in order to intensify the effects, an idea that may yet prove valuable. In the analog two tuning forks are provided, one at the sending and the other at the receiving station, each having attached to its lower prong a piston fitting in a cylinder. The two cylinders communicate with a large elastic reservoir filled with an incompressible fluid. The vibrations transmitted to either of the tuning forks excite them by resonance and, through electrical contacts or otherwise, bring about the desired result. This, I may say, was not a mere mechanical illustration, but a simple representation of my apparatus for submarine signaling perfected by me in 1892, but not appreciated at that time, although more efficient than the instruments now in use.

The electric diagram in Fig. 7, which was reproduced from my lecture, was meant only for the exposition of the principle. The arrangement,

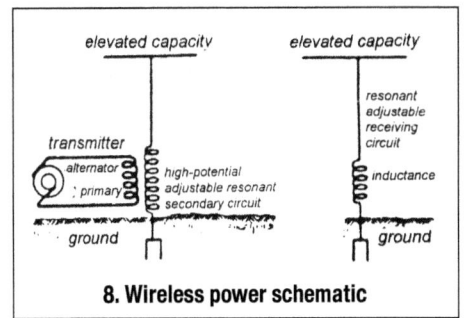

8. Wireless power schematic

as I described it in detail, is shown in Fig. 8. In this case an alternator energizes the primary of a transformer, the high-tension secondary of which is connected to the ground and an elevated capacity and tuned to the impressed oscillations.

The receiving circuit consists of an inductance connected to the ground and to an elevated terminal without break and is resonantly responsive to the transmitted oscillations. A specific form of receiving device was not mentioned, but I had in mind to transform the received currents and thus make their volume and tension suitable for any purpose. This, in substance, is the system of today, and I am not aware of a single authenticated instance of successful transmission at considerable distance by different instrumentalities.

It might perhaps be clear to those who have perused my first description of these improvements that, besides making known new and efficient types of apparatus, I gave the world a wireless system of potentialities far beyond anything before conceived. I made explicit and repeated statements that I contemplated transmission, absolutely unlimited as to terrestrial distance and amount of energy. But though I have overcome all obstacles which seemed in the beginning unsurmountable and found elegant solutions of all the problems which confronted me, experts are still blind to the possibilities which are within easy attainment.

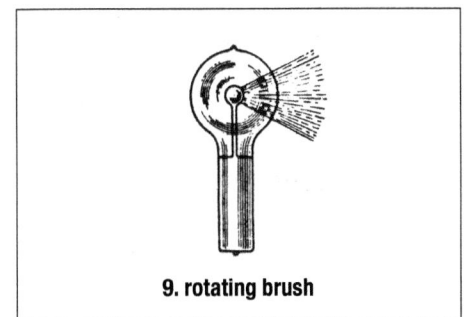

9. rotating brush

My confidence that a signal could be easily flashed around the globe was strengthened through the discovery of the "rotating brush," a wonderful phenomenon which I have fully described in my address before the Institution of Electrical Engineers, London, in 1892, and which is illustrated in Fig. 9. This is undoubtedly the most delicate wireless detector known, but for a long time it was hard to produce and to maintain such a sensitive state. These difficulties do not exist now, and I am looking to

wireless transmitter — adjustable inductance — primary condenser — wireless receiver

source of impulses of arbitrary frequency

circuit controller in synch with impulses

adjustable inductance

tuned secondary with adjustable capacity

primary circuit

the four tuned circuits of above diagram shown separately

energizing circuit 1 supplying oscillations of arbitrary frequency

adjustable inductance

transforming circuit 2 tuned to frequency of circuit 1 or harmonic of same

adjustable inductance

discharge circuit 3 similarly tuned

transforming circuit 4, similarly tuned

adjustable capacity

the corresponding four tuned circuits of the wireless system

energizing wireless primary 1 supplying oscillations of arbitrary frequency

wireless transforming circuit 2 tuned to frequency of circuit 1 or or harmonic of same

adjustable inductance

wireless receiving circuit 3 similarly tuned

adjustable inductance

wireless transforming circuit 4 similarly tuned

adjustable capacity

10. multiple tuned circuits   **("concatenated")**

Tesla's four-circuit tuned wireless system

Hertz wave system

11. Tesla's 4-circuit system and Hertz's

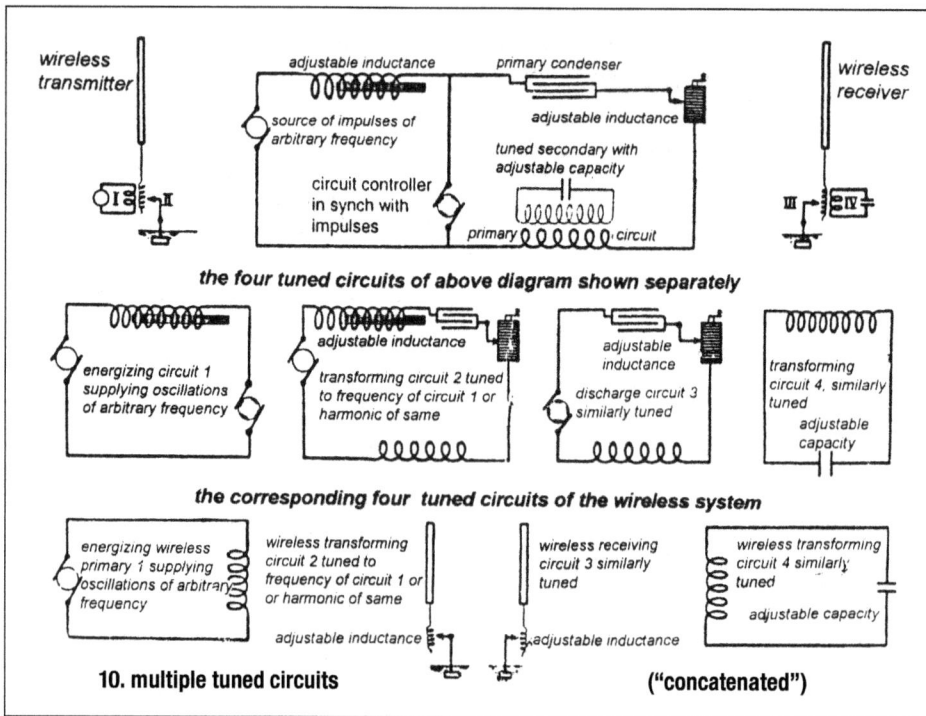

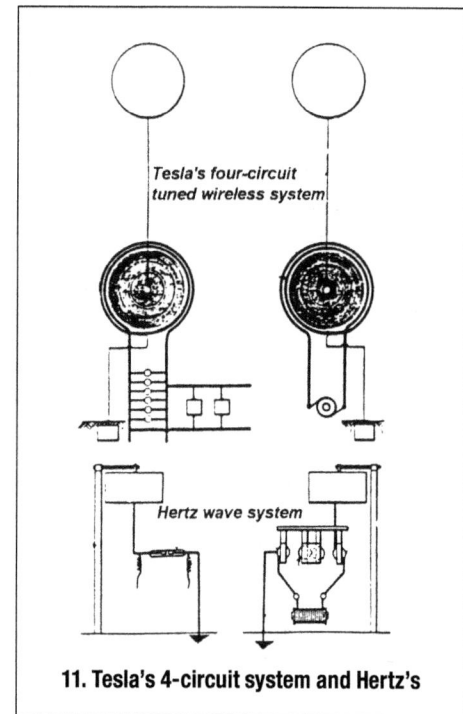

valuable applications of this device, particularly in connection with a high-speed photographic method, which I suggested in wireless, as well as in wire, transmission.

Possibly the most important advances during the following three or four years were my system of concatenated tuned circuits and methods of regulation, now universally adopted. The intimate bearing of these inventions on the development of the wireless art will appear from Fig. 10, which illustrates an arrangement described in my U.S. Patent No. 568,178 of September 22, 1896, and corresponding depositions of wireless apparatus. The captions of individual diagrams are thought sufficiently explicit to dispense with further comment. I will merely remark that in this early record, in addition to indicating how any number of resonant circuits may be linked and regulated, I have shown the advantage of the proper timing of primary impulses and use of harmonics.

enormously magnified

In a farcical wireless suit in London, some engineers, reckless of their reputation, have claimed that my circuits were not at all attuned; they asserted that I had looked upon resonance as a wild and untamable beast.

It will be of interest to compare my system as first described in a Belgian patent of 1897 with the Hertz wave system of that period. The significant differences between them will be observed at a glance. The first enables us to transmit economically energy to any distance and is of inestimable value; the latter is capable of a radius of only a few miles and is worthless.

In the first there are no spark gaps, and actions are enormously magnified by resonance. In both transmitter and receiver the currents are transformed and rendered more effective and suitable for the operation of any desired device.

Properly constructed, my system is safe against static and other interference, and the amount of energy which may be transmitted is billions of times greater than with the Hertzian, which has none of these virtues, has never been used successfully, and of which no trace can be found at present.

A well advertised expert gave out a statement in 1899 that my apparatus did not work and that it would take 200 years before a message would be flashed across the Atlantic, and he even accepted stolidly my congratulations on a supposed great feat. But subsequent examination of the records showed that my devices were secretly used all the time, and ever since I learned of this I have treated these Borgia-Medici methods with the contempt in which they are held by all fair-minded men.

The wholesale appropriation of my inventions was, however, not without a diverting side. As an example to the point, I may mention my oscillation transformer operating with an air gap. This was in turn replaced by a carbon arc, a quenched gap, an atmosphere of hydrogen, argon or helium, by a mechanical break with oppositely rotating members, a mercury interrupter, some kind of a vacuum bulb, and by such tours de force as many new "systems" have been produced. I refer to this, of course, without the slightest ill-feeling, let us advance by all means. But I cannot help thinking how much better it would have been if the ingenious men, who have originated these "'systems" had invented something of their own instead of depending on me altogether.

Before 1900 two most valuable improvements were made. One of these was my individualized system with transmitters emitting a wave-complex and receivers comprising separate tuned elements cooperatively associated. The underlying principle can be explained in a few words.

Suppose that there are n simple vibrations suitable for use in wireless transmission, the probability that any one will be struck by an extraneous disturbance is 1/n. There will then remain vibrations, and the chance that one of these will be excited is

$$\frac{1}{n-1}$$

hence the probability that two tunes would be struck at the same time is

$$\frac{1}{n(n-1)(n-2)}$$

and so on. It will be readily seen that in this manner any desired degree of safety against statics or other kind of disturbance can be attained providing the receiving apparatus is so designed that its operation is possible through the joint action of all the tuned elements. This was a difficult problem which I have successfully solved so that now any desired number of simultaneous messages is practicable in the transmission through earth as well as through artificial conductors.

The other invention, of still greater importance, is a peculiar oscillator enabling the transmission of energy without wires in any quantity that may ever be required for industrial use, to any distance, and with very high economy. It was the outcome of years of systematic study and investigation, and wonders will be achieved by its means.

The prevailing misconception of the mechanism involved in the wireless transmission has been responsible for various unwarranted announcements which have misled the public and worked harm. By keeping steadily in mind that the transmission through the earth is in every respect identical to that through a straight wire, one will gain a clear understanding of the phenomena and will be able to judge correctly the merits of a new scheme.

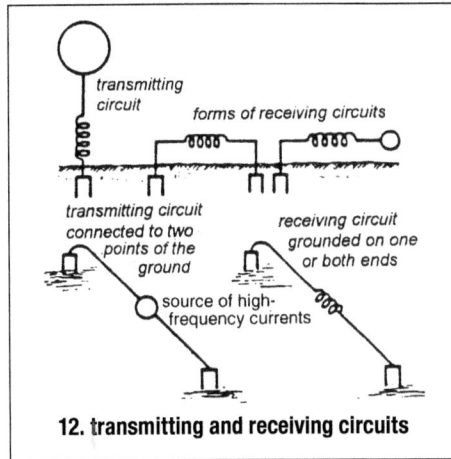

12. transmitting and receiving circuits

Without wishing to detract from the value of any plan that has been put forward, I may say that they are devoid of novelty. So for instance in Fig. 12 arrangements of transmitting and receiving circuits are illustrated, which I have described in my U.S. Patent No. 613,809 of November 8, 1898 on a Method of and Apparatus for Controlling Mechanism of Moving Vessels or Vehicles, and which have been recently dished up as original discoveries. In other patents and technical publications, I have suggested conductors in the ground as one of the obvious modifications indicated in Fig. 5.

For the same reason the statics are still the bane of the wireless. There is about as much virtue in the remedies recently proposed as in hair restorers. A small and compact apparatus has been produced which does away entirely with this trouble, at least in plants suitably remodeled.

Nothing is more important in the present phase of development of the wireless art than to dispose of the dominating erroneous ideas. With this object I shall advance a few arguments based on my own observations which prove that Hertz waves have little to do with the results obtained even at small distances.

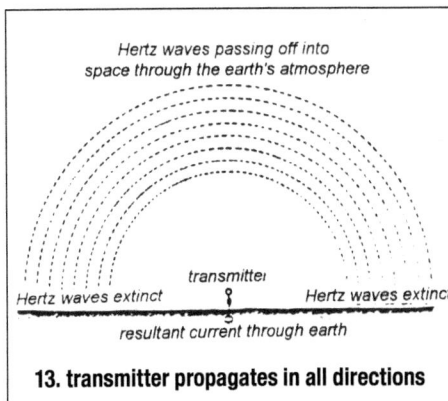

13. transmitter propagates in all directions

In Fig. 13 a transmitter is shown radiating space waves of considerable frequency. It is generally believed that these waves pass along the earth's surface and thus affect the receivers. I can hardly think of anything more improbable than this "gliding wave" theory and the conception of the "guided wireless" which are contrary to all laws of action and reaction. Why should these disturbances cling to a conductor where they are counteracted by induced currents, when they can propagate in all other directions unimpeded?

The fact is that the radiations of the transmitter passing along the earth's surface are soon extinguished, the height of the inactive zone indicated in the diagram, being some function of the wave length, the bulk of the waves traversing freely the atmosphere. Terrestrial phenomena which I have noted conclusively show that there is no Heavyside layer, or, if it exists, it is of no effect. It would certainly be unfortunate if the human race were thus imprisoned and forever without power to reach out into the depths of space.

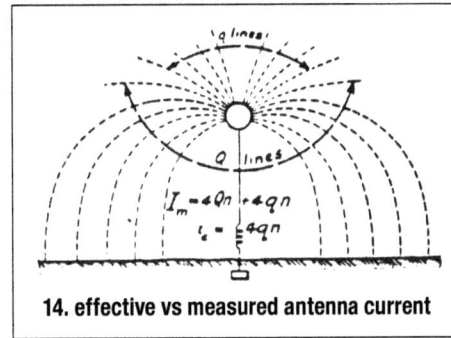

14. effective vs measured antenna current

The actions at a distance cannot be proportionate to the height of the antenna and the current in same. I shall endeavor to make this clear by reference to the diagram in Fig. 14. The elevated terminal charged to a high potential induces an equal and opposite charge in the earth, and there are Q lines, giving an average current =4Qn, which circulates locally and is useless except that it adds to the momentum. A relatively small number of lines q, however, go off to a great distance and to these corresponds a mean current of ie=4qn, which is due to action at a distance. The total average current in an antenna is thus Im=4qn, and its intensity is no criterion for performance. The electrical efficiency of the antenna is

$$\frac{Q}{Q + q}$$

and this is often a very small fraction.

Dr. L.W. Austin and Mr. J.L. Hogan have made quantitative measurements which are valuable, but far from supporting the Hertz wave theory they are evidences in disproval of the same, as will be easily perceived by taking the above facts into consideration. Dr. Austin's researches are especially useful and instructive, and I regret that I cannot agree with him on this subject. I do not think if his receiver was

affected by Hertz waves he could ever establish such relations as he has found, but he would be likely to reach these results if the Hertz waves were in large part eliminated. At great distance the space waves and the current waves are of equal energy, the former being merely an accompanying manifestation of the latter in accordance with the fundamental teachings of Maxwell.

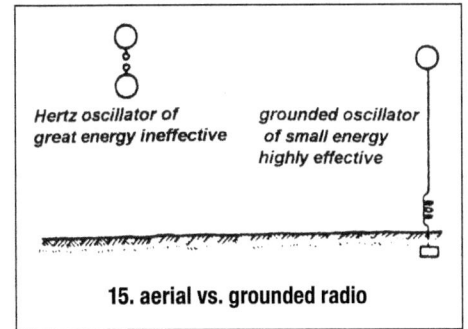

15. aerial vs. grounded radio

It occurs to me to ask the question: why have waves been reduced from the original frequencies to those I have advocated for my system, when, in so doing, the activity of the transmitting apparatus has been reduced a billion fold? I can invite any expert to perform an experiment such as illustrated in Fig. 15, which shows the classical Hertz oscillator and my grounded transmitting circuit.

It is a fact which I have demonstrated that, although we may have in the Hertz oscillator an activity thousands of times greater, the effect on the receiver is not to be compared to that of the grounded circuit. This shows that in the transmission from an airplane we are merely working through a condenser, the capacity of which is a function of a logarithmic ratio between the length of the conductor and the distance from the ground. The receiver is affected in exactly the same manner as an ordinary transmitter, the only difference being that there is a certain modification of the action which can be predetermined from the electrical constants. It is not at all difficult to maintain communication between an airplane and a station on the ground, on the contrary, the feat is very easy.

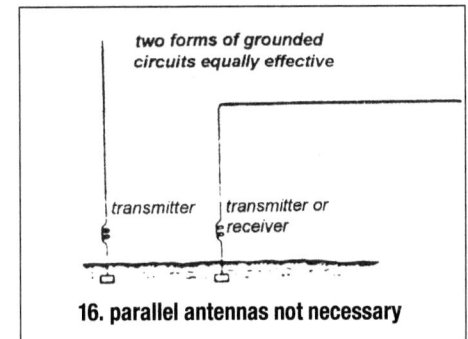

16. parallel antennas not necessary

To mention another experiment in support of my view, I may refer to Fig. 16, in which two grounded circuits are shown excited by oscillations of the Hertzian order. It will be found that the antennas can be put out of parallelism without noticeable change in the action on the receiver,

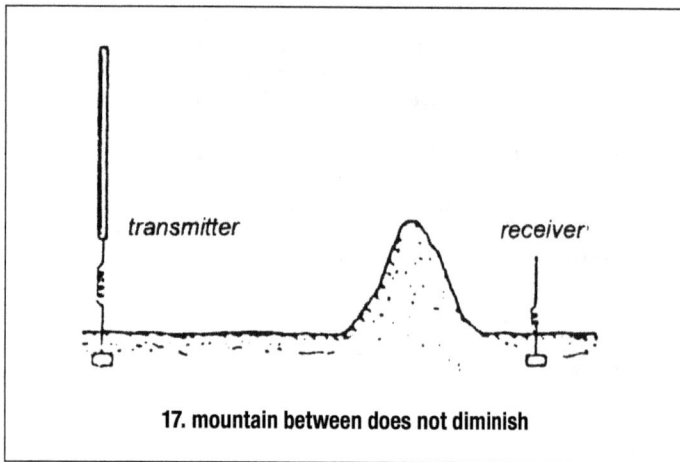

17. mountain between does not diminish

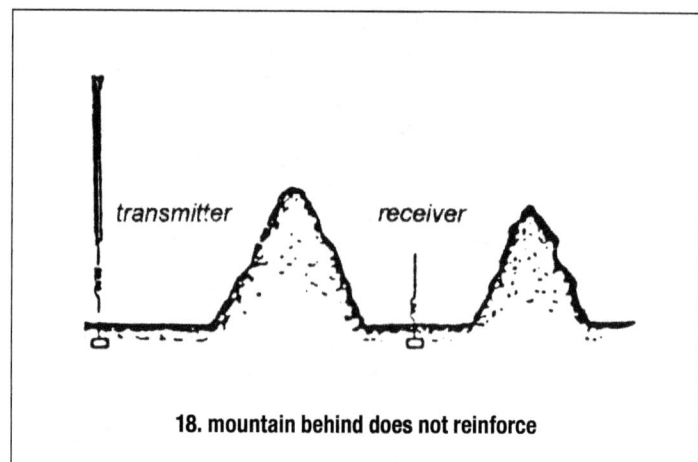

18. mountain behind does not reinforce

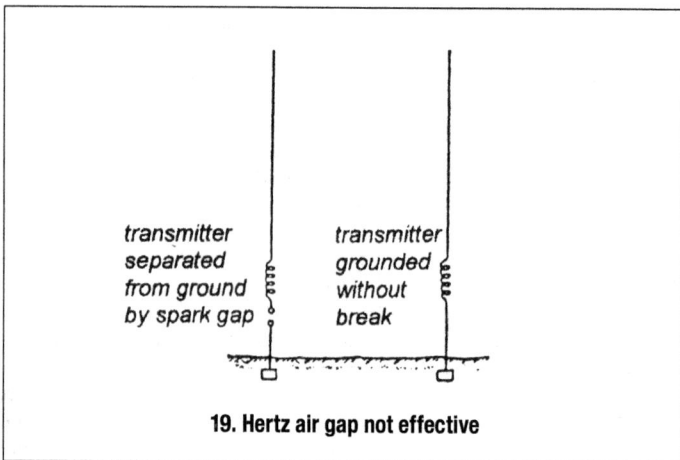

19. Hertz air gap not effective

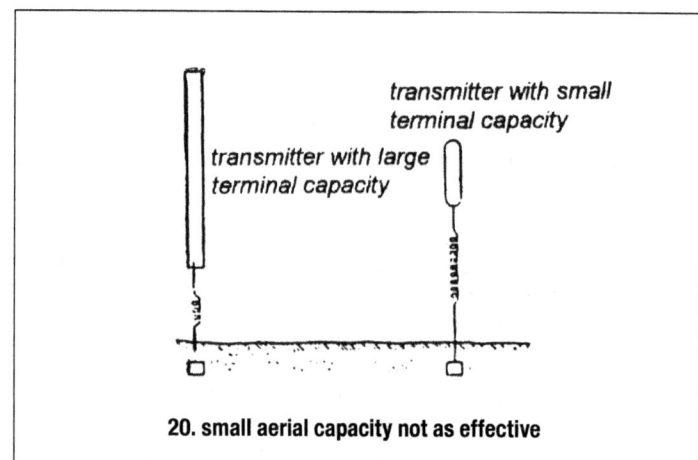

20. small aerial capacity not as effective

this proving that it is due to currents propagated through the ground and not to space waves.

Particularly significant are the results obtained in cases illustrated in Figures 17 and 18. In the former an obstacle is shown in the path of the Q waves, but, unless the receiver is within the effective electrostatic influence of the mountain range, the signals are not appreciably weakened by the presence of the latter, because the currents pass under it and excite the circuit in the same way as if it were attached to an energized wire.

If, as in Fig. 18, a second range happens to be beyond the receiver, it could only strengthen the Hertz wave effect by reflection, but as a matter of fact it detracts greatly from the intensity of the impulses because the electric niveau between the mountains is raised, as I have explained in connection with my lightning protector in the *Experimenter* of February, 1919.

Again in Fig. 19, two transmitting circuits, one grounded directly and the other through an air gap, are shown. It is a common observation that the former is far more effective, which could not be the case in a transmission with Hertz radiations. In like manner if two grounded circuits are observed from day to day, the effect is found to increase with the dampness of the ground, and for the same reason also the transmission through sea water is more efficient.

An illuminating experiment is indicated in Fig. 20, in which two grounded transmitters are shown, one with a large and the other with a small terminal capacity. Suppose that the latter be one-tenth of the former but that it is charged to ten times the potential, and let the frequency of the two circuits and therefore the currents in both antennas be exactly the same. The circuit with the smaller capacity will then have ten times the energy of the other, but the effects on

the receiver will be in no wise proportionate.

The same conclusions will be reached by transmitting and receiving circuits with wires buried underground. In each case the actions carefully investigated will be found to be due to earth currents. Numerous other proofs might be cited which can easily be verified. So for example oscillations of low frequency are ever so much more effective in transmission, which is inconsistent with the prevailing idea. My observations in 1900 and the recent transmissions of signals to very great distances are another emphatic disproval.

The Hertz wave theory of wireless transmission may be kept up for a while, but I do not hesitate to say that in a short time it will be recognized as one of the most remarkable and inexplicable aberrations of the scientific mind which has ever been recorded in history.

Explanatory Notes

by George Trinkaus

oscillation transformer, pg. 67. i.e., the Tesla coil, which became Tesla's potent transmitter.

Maxwell's electromagnetic theory, pg. 67. James Clerk Maxwell (1831-79) developed the mathematics of electromagnetic phenomena and predicted that oscillations would produce wave propagations via an etheric medium, causing actions at a distance (i.e., radio). Maxwell identified the phenomenon with the physics of light, starting a tradition Tesla was to dispute.

Hertz's results, pg. 67. Heinrich Hertz (1857-94) conducted experiments with a spark-gap oscillator that appeared to confirm Maxwell's mathematical predictions of radio propagation. He, too, contended that radio behaved like light-rays, having experimentally demonstrated reflection, etc. His ideas became fixtures in radio theory, and Hertz has been honored in recent years by renaming the unit of frequency after him (hz), although there is no law against using instead the old descriptive cycles per second (cps). Tesla contended that the waves Hertz described were not those that made radio work.

Rhumkorf coil, pg. 67. A type of laboratory high-voltage induction coil that has a built-in vibrator enabling it to run on DC.

1900 wireless transmitter, pg. 67. i.e., the magnifying transmitter, which Tesla developed during his Colorado Springs adventure of 1899-1900. Evolved from the Tesla coil and employing an extra third coil, this became Tesla's ultimate tool for radio and wireless power transmission.

heliographic, pg. 68. An old method of signaling that used a movable mirror to reflect beams of sunlight to a distance.

alternator, pp. 68 and 69. An AC dynamo cited by Tesla here as his source of high-frequency oscillations when he lectured on the single-wire-without-return phenomenon. Tesla is on record as achieving 30 kc with such a device, which he also employed for experiments in high-frequency lighting. Tesla held many dynamo patents and was the consummate engineer in the field, but he abandoned the device because of its frequency limitations, employing instead his Tesla coil. The alternator did have a role in early low-frequency radio transmitting.

article in Century Magazine, pg. 69. Tesla's article "On Increasing Human Energy" of June 1900, following his return from Colorado, was celebrated when it appeared, although, as an account of Tesla's Colorado work, it had been drastically edited.

rotating brush, pg. 69. A glowing discharge formed when high-voltage, high-frequency currents are applied to a terminal centered in a high-vacuum bulb. The discharge resolves into a ray so sensitive to any etheric disturbances in its vicinity that it responds to a person approaching. Tesla expresses enthusiasm here and in other writings for its potential as a radio detector. Exactly how its sensitivity could be harnessed for this purpose is not clear, but in the same paragraph above he refers to a high-speed photographic method, and a year later in the *Electrical Review*, he refers to a similar device and to "a vacuum tube of high intensity which throws its light through a fine slot on a sensitive plate." In the same article he progresses to speculations on television, using that word, and says that this is a "subject to which I have devoted

more than 25 years of close study." Imagine a Tesla TV in which a rotating-brush device both detects the signal and paints the picture in one simple, direct translation. Is this what he had in mind?

concatenated, tuned circuits, pg 70. Linked together, as in a chain. This basic invention of Tesla's was adopted by modem radio as indispensable, while his other basic precepts were ignored. Tesla invented more elaborate inter-linkings, such as multiplexing.

farcical wireless suit pg. 70. Probably Marconi Wireless Telegraph vs. the United States. Marconi patented Tesla's technology then challenged one of Tesla's fundamental patents (645,576, Fig. II). Marconi had the wealth and clout to keep the farce going until 1943, when it was finally struck down by the U.S. Supreme court.

Borgia-Medici methods, pg. 70. Ruthless and devious, was were some of the stratagems of these two ruling families of 16th-century Italy. Machiavellian.

Heavyside layer pg. 69. 1. The ionosphere. 2. The radio-reflective layer presumed to exist within it.

Parallelism (of antennas), pg 71. Tesla challenged the Hertzian notion that transmitting and receiving antennas must be similarly oriented, or polarized, i.e., vertical-to-vertical or horizontal-to-horizontal. (In the vertically oriented world of CB, I have used horizontal antennas with success in both mobile and fixed situations.)

Electric niveau, pg. 72. (French) the amount of natural electricity distributed over the land.

Part 3
Tesla DIY

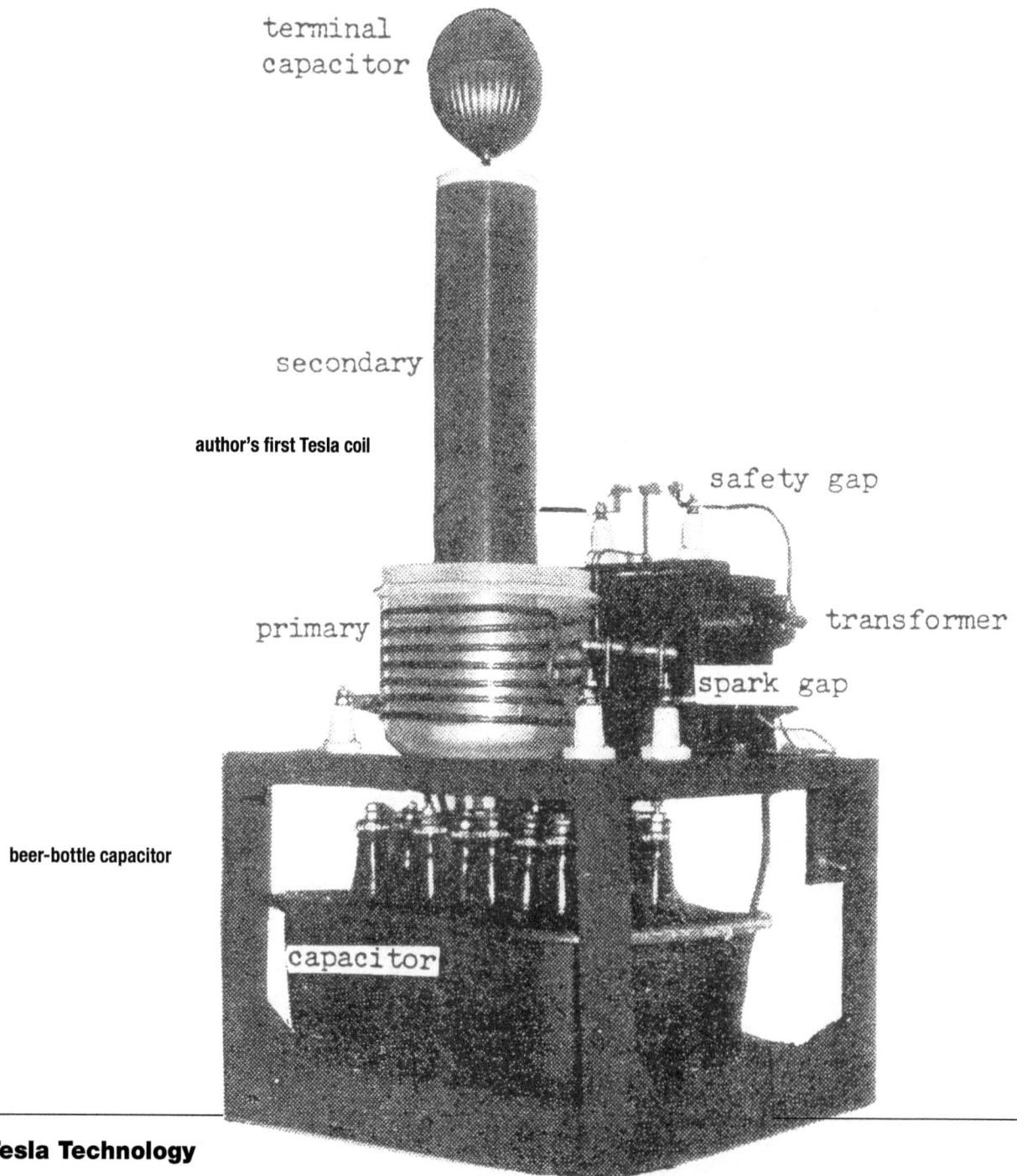

terminal
capacitor

secondary

author's first Tesla coil

safety gap

primary

transformer

spark gap

beer-bottle capacitor

capacitor

CHAPTER 1

Build a Tesla Coil

I have built a few of the Tesla coils shown, including the "recipe" coil below. I also illustrate for inspiration many other devices from the history of Tesla-coil building, including some by Tesla himself. In either case, you are on your own in the agonies of DIY. So don't call me for advice. If the information you need is not here, go digging.

The secondary coil

This is a good place to start. Its size will determine the scale of the whole project, and building it is one of the more patience-demanding tasks and best gotten done first. For the secondary coil form, you will need a length of plastic tubing or some other cylinder of insulating material. Before plastics, cardboard tubing was often used. This has to be dried slowly in an oven and sealed with several coats of shellac or varnish. Because of its easy availability in building-supply and hardware store, many Tesla-coil builders now use PVC tubing for coil forms, and PVC does work, particularly the thin-walled variety. Superior insulating (dielectric) qualities are to be found, though, in tubing of acrylic (the thinner the better), of phenolic, and in the so-called air-wound coils. "Air-wound" means the wire is held only by slender insulating supports. Air-wound is regarded as dielectrically superior. The less form, the better. `

Most home-built coils use long narrow secondaries because of the availability of small diameter tubing, but Tesla's patents show wide drum-like secondaries with length-to-width ratios on the order of two to one. This tradition is honored more in the "second generation" of Tesla-coil building. See next chapter. In that parlance, we are building here a "first generation" Tesla coil.

On your form you will wind something under 500 turns typically. Use a number 28 up to number 24 solid insulated wire: enamel-insulated magnet wire, or even the more widely available plastic-insulated wire. You can wind a secondary of an arbitrary number under 500 turns, or calculate more precisely if you want a particular frequency.

operating frequency

You can ball-park the frequency that the secondary will favor by using Tesla's quarter-wave formula. It says that the length of the secondary wound up should be one quarter of the wavelength, or an odd multiple of that number. Decide upon your frequency in megahertz (MHz), and divide this into 246. This will give you the quarter-wavelength in feet. Find an odd number or fraction you can multiply this answer by that will give you the length of

author's winding jig

wire to wind on your form. This exercise is simplistic since something called the capacitive reactance of the coil comes into play. The coil will also want to vibrate at various multiples (harmonics) of its fundamental frequency. The dimensions of your terminal, will also determine the frequency of the completed Tesla coil to a great degree. The capacity of the terminal becomes a major reactive element in the system, which can be tuned by designing a terminal of appropriate dimensions.

The quarter-wave exercise is rough, but it will at least give you something to go by. With mathematical procedures, you can dimension a Tesla coil for specific primary and secondary frequencies, for voltage input, and for harmonious scaling of components. In practice, most of us build with the components we can easily get a hold of, some rules of thumb, and trial and error.

Because a Tesla-coil circuit is like a radio tuning tank, the Wavelength, Frequency and Oscillation Constant tables on page 104 in "Building Radio Tanks," can be helpful in designing for frequency range. For a particular frequency the table shows the oscillation constant, which is an index of the reactance of the capacitor and primary inductance combined (capacity in mfd. x inductance in microhenries = oscillation constant).

You can now calculate the length of the tubing you'll need. Wind an inch's worth of wire on the tube. Count the turns, and you have the number of turns per inch. Use the table shown on page 86. Allow for an extra half inch of tubing on either side of the winding.

To wind the coil, drill a hole in the tube a half inch from the bottom, tape the wire to the inside, pass it through, and proceed to wind. I know of no advantage to winding clockwise as opposed to counterclockwise, but there may be one. Many builders contrive some sort of jig to wind the coil on. It is possible, though, to wind the coil hand-held, but it helps to have a friend hold the spool and dispense the wire as needed. The trick in any method is to keep the wire under constant tension with thumb pressure as it is being

a space-wound coil

just bundles of insulated wire coiled at the foot of the secondary. A single turn works in some cases. Experiment for best results. Design will determine the electrical coupling of primary to secondary, an important factor. You can use bare wire or make elegant coils out of copper tubing. Either solid or stranded conductors can be used. Whatever you use; it should be highly conductive, offering negligible resistance to the pulsating current. Wind the primary in the same direction as the secondary.

Tesla preferred copper ribbon, and since I came across some heavy enamel-insulated ribbon at an electronics surplus store, I tried it out. It is shown in the illustration above. It looked impressive, but was very difficult to shape, and in an early trial some hot electric bolts shot up out of the thing, sliced through the plastic of the coil form, and pierced the insulation on the secondary. (This trauma was the result of too many beer bottles, that is, too much capacity.) Eventually I went to a heavy, #8 stranded, plastic-insulated battery wire from the auto store, where you can get it as heavy as #4.

Some experimenters have used flat spirals and even cone shapes that open upward, but most primaries are cylindrical. The coil form, if you use one, should be at least twice the diameter of the secondary. This raises a problem, since you can't easily find PVC or other insulating tubes in these large diameters. Look around for any cylindrical something that meets your requirements. It is ideal to experiment with different diameters. I found my primary form on the housewares shelves at the supermarket (see below). Some builders with carpentry skills construct special structures of upright dowels for air-wound coil forms. If your solution is in wood, use a well-dried, well-sealed hardwood.

One old rule of thumb says the primary and the secondary should have equal weights of copper.

The primary can have a role in fine-tuning the Tesla coil. Turns can be added or removed. Many primaries have a tap in the form of a clip on the wire from the capacitor that can be moved from turn to turn. Small sections of insulation can be removed from a primary winding in order to accommodate the clip. These sections should be removed in a stepping pattern so they are not opposite one another at adjacent turns. (Tesla often used a separate variable inductor coil to tune his Tesla-coil circuits.) Some Tesla coils have a secondary that can be moved up and down in respect to the primary. Both these methods tune by changing the "coupling" of the coils. A well-proportioned primary should have a height (vertical distance from the first turn to the last) that measures about one-half the primary's diameter.

wound on the form. I built the winding jig shown in the photo using 3/8-inch steel all-thread, 24" long as the winding shaft and 5/8-inch wooden dowel as the wire-feed shaft, The handle is a VW window crank. (Shown later is the winding jig the author built for his "third-generation" Tesla coil, which has a shorter bed.)

Some builders turn a groove on the coil form with a lathe along its length so there is extra space between turns, thus assuring minimal bleed-over and capacitive effects between turns. An easier turn-spacing technique is to wind string or monofilament fishing line alongside the wire. Some builders put extra space between the turns only at the high-potential top end of the coil for the last 10 to 15 percent of the winding.

Drill another hole at the end of the winding and fasten the wire inside the form temporarily with a piece of tape. After winding, spray the completed coil with several coats of some plastic coating. This helps keep the turns in place, protects the winding, and further insulates it. A number of light coats helps to eliminate the possibility of air pockets in the coil. These could cause operating problems, to the extent that Tesla often resorted to winding his coils under oil.

recipe secondary coil

Here are the specs for the secondary on my Tesla coil shown in the photo: Coil form is PVC with a 3-1/2 inch outside diameter and 18 inches long. The winding, on 17 inches of the form, is 437 feet of #28 solid, plastic-insulated wire, 477 turns.

the primary coil

The primary coil is just a half dozen or so turns (rarely more than ten) of some very heavy conductor. Standards here are loose. Almost anything seems to work. I've seen primaries that were

recipe primary coil

My supermarket coil form is a Rubbermaid "Servin' Saver" 10-cup cylinder. The diameter is 7 inches, height 5-3/4 inches. The top can be cut open for the secondary to pass through, but I just discarded it. Onto my coil form are wound the six turns

If Joey can, you can.

of plastic-insulated #8. The insulation was removed in staggered one-half-inch segments to accommodate an alligator clip.

the transformer

The transformer is one component that you will have to shop around for. Needed is one that can deliver something between 6,000 and 15,000 volts, at 30 to 60 MA. Don't despair; these transformers are everywhere, although you may not be aware of them. Neon signs (a Tesla invention) are driven by them. Oil furnaces are ignited by them.

The neon is superior to the oil-furnace ignition transformer, being made for continuous operation. Neon sign transformers put out 3,000 to 15,000 volts, 30 or 60 MA, with power ratings up to 1800 VA, builders preferring the upper range. (Oil furnace ignition transformers usually run around 10,000 volts.) You can get a neon transformer, new or used, at neon sign shops or at surplus electronics outlets or junk stores, or at wrecking or metals recycling yards (the best bargains). Two manufacturers to do a search on are France and Allanson.

Neon transformers are packed in an insulating, weather-proofing tar and often fail because carbon tracks can form in this tar and cause a short circuit. Some manufacturers offer an open-core version without the tar or case, but only up to 7500 volts. I know one builder who melts out this tar with potent solvents, but this is an unappealing job.

Transformers can be ganged up in parallel to power big Tesla coils. I've heard of as many as eight hooked together to deliver two kilowatts. Some builders wind their own transformers. Specs can be found in electrical texts. Transformer outputs can be rectified. This serves up DC for the capacitor to oscillate with, mitigating somewhat the influence of the transformer's 60-cycle vibration. One high-voltage rectifier is the ECG-513. These 3-3/4-inch heavy duty diodes are rated 45 thousand volts. The DC is a particular shock hazard at these voltages.

Also, I would hesitate to apply DC to a salt-water capacitor because electrolysis might occur, so there would be off-gassing of explosive hydrogen.

battery systems

A battery powered Tesla coil can have great practical value. In remote areas where there is no electrical service, batteries can be charged by solar-electric or wind power and used to drive a Tesla-coil powered high-frequency lighting system. A simple method is to power the transformer with a power inverter.

Tesla coils designed specifically for 12-volt battery power can use the automobile ignition coil as a transformer. Of course, the DC Must be made to pulsate to drive an ignition coil. That is, you need something to do the job ignition points do in a car. An electro-mechanical vibrator can be specially built. As we will see, solid-state circuits can do the job, and by using integrated circuits (555 Timers, pulse-width modulators) it is possible to control the primary frequency and wave shape, a real luxury. Old-time builders of battery coils used laboratory induction coils or the then-handy Ford spark coil, which had a built-in adjust-able vibrator. A few of these are still around. You'll need one that gives a 3/4 to 1 inch spark.

flyback transformers

Battery Tesla coils can be made from old cathode-ray-TV flyback transformers, if you can find one. Using only two transistors and two resistors in an inverter circuit, a flyback transformer becomes a little Tesla coil in itself.

The flyback is a multilayer coil. It generates the high voltage for the picture tube and is a little Tesla coil in itself, demon-strating that when pushed by necessity conventional engineering does resort to Tesla technology. The flyback is a 12-volt device.

Allanson neon transformer

author's fly-back Tesla coil

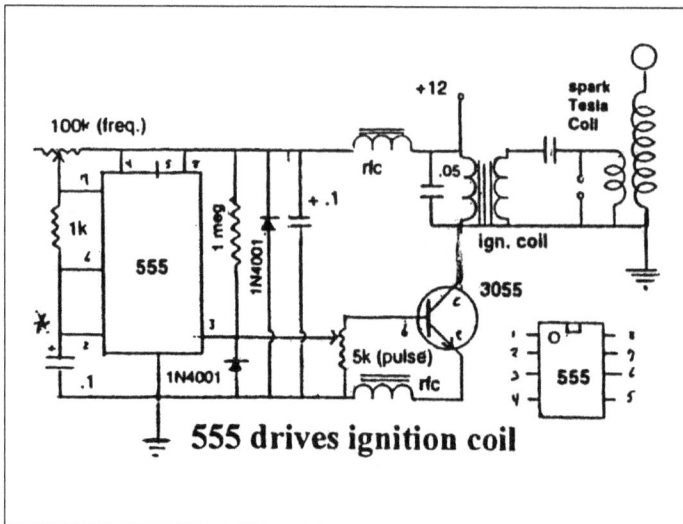

555 drives ignition coil

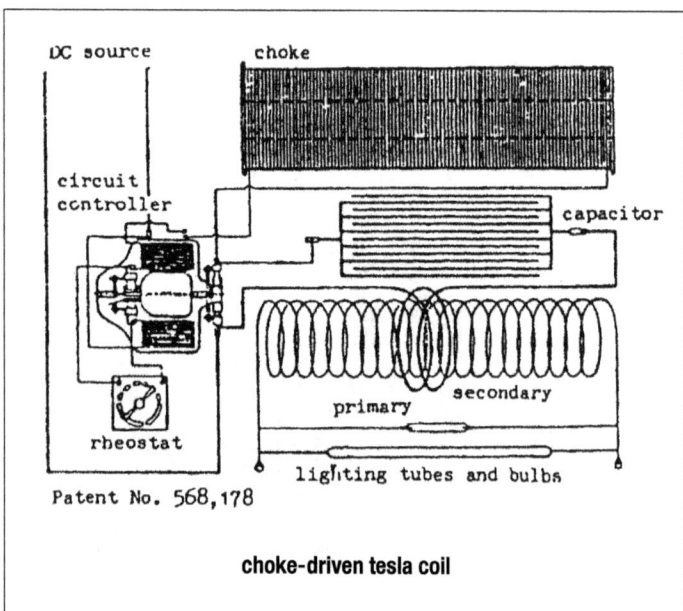

Patent No. 568,178

choke-driven tesla coil

A limit of the flyback is that it cannot be pushed to resonance without burning out. The same chopper that drives the ignition coil can also drive a flyback. However, the most frequently seen flyback project, including mine above, is driven by an inefficient inverter circuit.

chopping battery power

Before we can put battery power to work for any alternating system like a Tesla coil, the battery direct current must be interrupted into a series of pulses, or "chopped." (The power inverter that converts battery DC to household AC for RVs, boats, and home power is a sophisticated type of chopper.)

Another example: a battery Tesla coil seen in the old literature uses a model-T spark coil for its high-voltage supply transformer. Like the laboratory induction coil, the model-T coil did its chopping by means of a doorbell-like magnetic interrupter. Later in the evolution of the automobile, chopping would be done by the ignition points, and later by a solid-state module. This kind of chopping is at low frequencies, down in the audio range, and at relatively low voltages. In high-voltage, high-frequency Tesla-coiling, we might construe the spark gap as a sort of chopper, be it a static gap or a rotary gap. Tesla's evolutions of the rotary spark gap, his "circuit controllers," might be seen as consummate mechanical choppers. These had frequency limitations which solid-state choppers promise to surpass.

a system using chokes

One of Tesla's patents shows a Tesla coil that runs on DC by means of a choke. A choke is a coil that uses electromagnetic effects to influence a circuit. Tesla's circuit here has no transformer. Instead the DC source charges a large choke (specs unknown). When the DC is switched off, the choke's magnetic field collapses, releasing suddenly a current much higher in potential than that from any battery. This is conducted to the capacitor. A motor-driven rotary switch links DC source to choke, choke to capacitor, and capacitor to primary at just the right instants, quite a challenge. I don't recommend this as a project.

In AC Tesla coils, chokes are also useful. Chokes oppose surges and can help reduce the problem of 60-cycle waves disturbing the rhythm of the primary circuit. Even small chokes placed in line with the transformer outputs will help protect the transformer from kickback.

safety gap

One of the big bummers plaguing Tesla coil experimenters is the accidental burning out of transformers. This is caused by high energy kick-back from the capacitor which fatally overloads the transformer's secondary or burns carbon tracks in the potting tar. Transformers can be easily protected by means of a safety gap like the one sitting on top of the transformer in the photo of the completed Tesla coil above. A safety gap is constructed like the main spark gap but has a center electrode which conducts overloads safely to a good earth ground. The safety gap should be adjusted to fire intermittently. With a safety gap and a pair of chokes, you can experiment freely with your coil and run it for long periods with peace of mind.

recipe transformer

The recipe transformer is an Allanson 12,000-volt neon sign transformer that I happened upon in a junk store (price: $20). The choking coils have 16 turns of number 10 insulated wire on 3/4 inch PVC. The safety gap is made from two brass angle brackets and three brass bolts, a piece of scrap brass strap and beautiful (although not absolutely necessary) porcelain stand-off insulators, which as you can see, adorn my Tesla coil elsewhere and which I lucked out on at an electronics surplus store.

The safety gap could just as well have been built with three brass angle brackets mounted on a slab of plastic or wood. If you want them, you can make your own stand-off insulators out of small glass or plastic bottles. Run a brass bolt up through the cap for a terminal. If it's a glass bottle, glue it to the chassis. If plastic, drill the bottom and mount it with a brass wood screw.

the capacitor

The capacitor is one component that many builders choose to buy these days, but, since commercial capacitors that can meet the high voltage requirements of the Tesla coil can be difficult to find (and may be expensive when you do find them) many builders want to build their own. However this can be a challenging project, and the result can never perform like the commercial capacitor. If you must go DIY, I recommend the recipe beer-bottle capacitor shown below.

Look for capacitors that alone or in combination give you capacities in these ranges: Small tabletop coils (on the order of my recipe coil and smaller): .001 to .01 mfd. Medium-sized coils: .02 to .03 mfd. Large coils: .05 mfd. and up. Look for high-voltage "doorknob" caps.

Capacity can be built up by connecting capacitors in parallel. So wired, capacities are additive:

$$C = C1 + C2 + C3, \text{etc.}$$

Your capacitor should handle at least the output of your transformer. The high-voltage capacitor is rated for puncture voltage. If you have a 12,000-volt transformer, you will have to find a capacitor rated 24,000 volts DC, for the capacitor must have twice the DC rating for AC usage. Since capacitors rated this high can be difficult to find, you will have to gang them up in series. But capacitors in series have a decreasing effect on total capacity:

$$1/C = 1/C1 + 1/C2 + 1/C3 \ldots$$

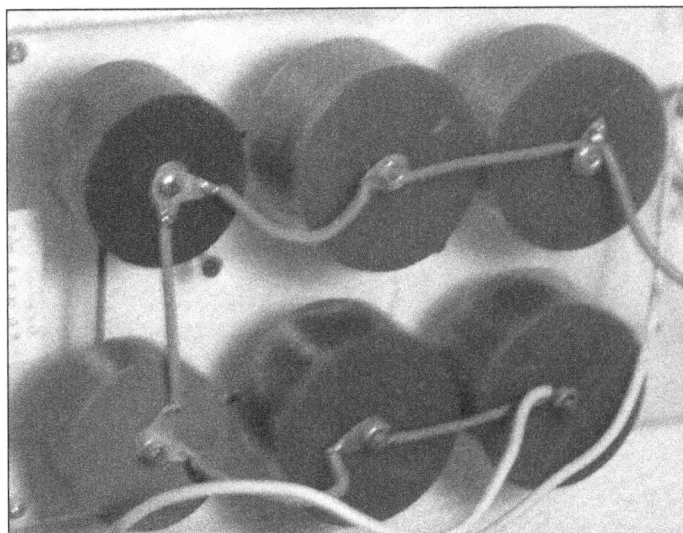

doorknobs

You must use high capacities to adjust for this loss. The best commercial types for this high-voltage, high-frequency work are pulse-discharge capacitors having polypropylene or polyethylene dielectrics, also mica, and ceramic, including the high-voltage caps called "doorknobs." Look for these in surplus electronics stores, and good luck.

The traditional home-built capacitor is made of sheets of metal foil interleaved with sheets of glass. The glass here is the dielectric and an effective one. The material used for the dielectric is critical to its operation. Some are far more effective than others. The effectiveness of a particular material used for a dielectric determines how large a capacitor must be to achieve a certain capacity. Effectiveness is expressed in a material's dielectric constant. Air, which breaks down pretty easily, serves as the standard and has a dielectric constant of 1. Transformer oil is 2.2, caster oil is 4.7, Formica is 4.8, mica is 6, glass is 7.8, and tantalum is a whopping 140. Polyethylene (2.2) is available in sheets, is lighter and easier to cut than glass, and will withstand high voltages. (1/32-inch holds up at 45 KV.) Polyethylene may make the glass-and-foil capacitor obsolete.

The higher the dielectric constant, the closer the plates can be to one another, and the closer the plates, the smaller the plate area needed to achieve a certain capacity. It's all in this formula which is good to have around if you're designing your own capacitor:

$$C = .224K \text{ A/d (n - 1).}$$

C is the capacity in picofarads (pf). Move this decimal six places to the left for mfd. K is the dielectric constant, A is the area of one plate in square inches, d is the distance between plates in inches, and n is the number of plates.

You can see how important the dielectric constant K is in this formula. An oil capacitor of ten 8"x 8" plates, 1/4" apart will give you 1135 pf or about .001 mfd., while glass would give you 4025 pf., or about .004 mfd., an increase by a factor of nearly

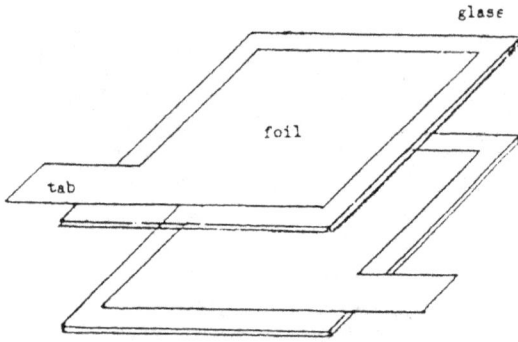

glass-and-foil capacitor construction

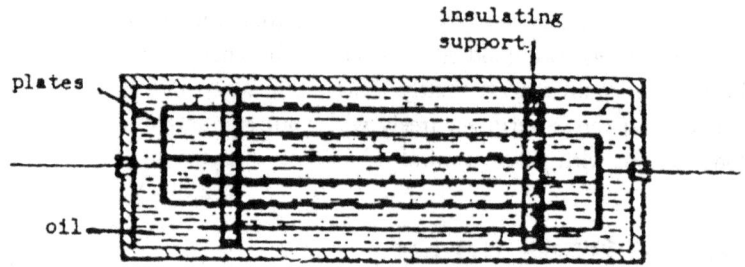

Patent No. 464,667

Tesla's oil capacitor

four. A capacitor with an oil dielectric would be awkwardly large and heavy, but effective and indestructible. Glass capacitors, though more effective ounce for ounce, can be broken through electrical as well as mechanical stresses.

glass and foil capacitor

An old tradition among builders, but my experience has not been so favorable that I would recommend the glass-and-foil. To make a glass-and-foil capacitor, you cut out rectangular sheets of aluminum foil, making them about two inches smaller in length and width than the glass plates. On one corner of the foil sheets leave a tab about 3 inches long and an inch or so wide. Spread a varnish on the glass and, while it is still wet, lay down the foil, leaving a one-inch margin of glass all around. Roll the foil flat with a dowel or something, rolling from the center, to remove all the air.

Arrange the tabs left-right, left-right, the idea being that tabs of alternate layers can be connected together. Smear some mineral oil on the one-inch margins left between foil edge and glass edge for added insulation. Bind the layers together with some insulated cord or ribbon. Attach a heavy wire to each set of tabs. Then drop the whole assembly into a plastic or wooden box filled with mineral oil. This will insure insulation. A polyethylene capacitor can be constructed on similar principles.

oil capacitor

To make an oil capacitor I refer you to Tesla's 1891 patent. Tesla in his patent says he has found solid dielectrics; like glass and mica to be "inferior" for his demanding uses, and that "highly efficient and reliable" capacitors can be made using oil as the dielectric. Oil dielectrics are self-repairing. An oil capacitor of sufficient rating will be large and cumbersome, as the dielectric constant of oil is relatively low. However, size and weight didn't worry Tesla, nor should they necessarily worry you. The plates in Tesla's oil capacitor can be of rigid metal. Aluminum sheet or plate will do. The plates are held in position by "strips of porous insulating material."

salt-water capacitor

None of the above capacitors offer much adjustability, although this is a feature that is very desirable when tuning the completed coil. In 1900, when Tesla left Manhattan to go to the wide-open spaces of Colorado Springs to build his magnifying transmitter, he reverted to a glass dielectric in the form of a bottle, and to salt-water, which served as the "plates" of an adjustable electrolytic capacitor.

Into a big galvanized tub of salt-water, Tesla set a bunch of large mineral-water bottles, which themselves contained salt-water. Salt-water is a conductive medium, an electrolyte. (Tesla had earlier patented a couple of electrolytic capacitors.) The salt-water in the bottles constituted one set of "plates", the salt-water in the tub the other set. The bottle glass was the dielectric. A connection was made to the tub, and each bottle had an electrode through its opening with a terminal on it. By connecting bottles in and out of this parallel circuit, Tesla could vary the total capacity by known increments.

recipe: beer-bottle capacitor

Taking a cue from Tesla, I constructed a salt-water capacitor on a smaller scale. Galvanized tubs are available at your Ace Hardware, but instead I bought at the supermarket a plastic dishpan and lined it with aluminum foil. The dishpan is from our friends at Rubbermaid and measures 11-1/2" x 13." A 1/4" x 1" brass bolt through a hole already in the dishpan's lip serves as a terminal for the foil. Into the dishpan fit nicely my sixteen Henry Weinhard ale bottles. The bottle should have the long-neck design with screw-on caps. The quality of the glass for electrical purposes may vary significantly from brand to brand, and I make no special claims for what I used. Tesla had his mineral water bottles sent from New York because he believed a particular brand's glass superior. Remove the labels.

Next you will need to devise electrodes and some sort of stopper to hold them in the beer bottles. For the electrodes I used 1/4-inch zinc-coated carriage bolts, six inches long. The threaded end goes up through the stopper and is attached by nuts on both sides, leaving some threading for the terminal (16 bolts, 48 nuts for the whole array.) The zinc-coated bolts will corrode in time. Stainless steel would be better and would be available from marine hardware sources.

The stopper for the electrode shown is fashioned from the bottle cap. Do not attempt to drill this flimsy metal; in fact it can be dangerous. Use a punch to make 1/4-inch holes in the screw-on caps for the electrodes to go through. (If you can't put your hands on a punch, use some other stopper method, perhaps rubber stoppers obtained from a supplier of lab equipment.) Smear some silicon sealant on the underside of the caps to prevent leakage.

In a suitable container, mix 3 gallons of strong solution of water and common table salt. (Sea salt is somewhat more conductive.) Fill each bottle to a depth of 4-3/4 inches. You now top off the bottles with an insulating layer of mineral oil. Fill the dishpan with saltwater to within about 1/4-inch of the top. Use some heavy conductor to interconnect the bottle terminals. Short pieces of stout (#10 to #6) wire with lugs crimped on the ends would be practical, but I made links by pounding flat some 1/4-inch copper tubing that was handy and punching out holes to fit over the terminals.

Assuming a K of 7 for the beer-bottle glass, each bottle is worth about .0005 mfd. The 16 together give about .008 mfd. This turns out to be more than necessary for the recipe coil, which never needed more than 6 to 10 bottles. The 16-beer-bottle capacitor has plenty of reserve for use with future larger coils. When you test, connect only a few of the bottles and work up in capacity as needed.

You may want to scale down to a more compact capacitor on the same principles. For example, the 10-ounce Kikkoman soy-sauce bottles have these perfect little plastic screw-on caps and pouring inserts to accommodate the electrodes.

capacitor problems and hazards

Even if you are using commercial capacitors, be aware that they are being stressed under high voltages and at high frequencies and may fail or become dangerous, whatever the voltage ratings. Dielectrics heat under high frequencies. (There is an industrial heating process called high-frequency dielectric heating.) Watch all capacitors for excessive heat. An oil capacitor may be unstable in frequency because of convection currents in the oil

Tesla's salt-water capacitors

3 nuts

twist-off cap

apply silicone sealant here

1/4" x 6" zinc-coated carriage bolt

electrode for beer-bottle capacitor

series gap

magnetic blow-out   electromagnet

mica

two Tesla gaps

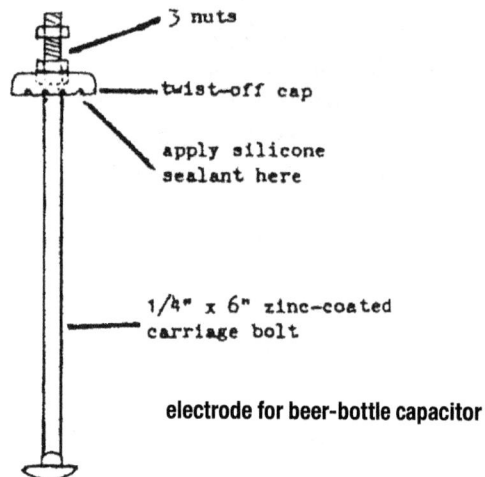

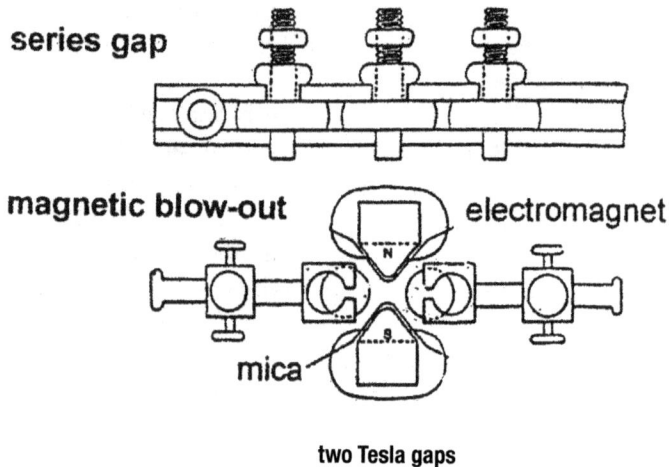

due to dielectric heating. Homemade capacitors should be closely observed for internal flashing. It's advisable to have see-through containers. Mineral oil can ignite and burn.

Look for flashing in a glass-and-foil capacitor and in the beer bottles, and shut down if you see it. The problem may be a bad connection or bubbles that need to be worked out of the fluid. One type of transmitting capacitor is in a glass cylinder containing an insulative oil having PCBs. Take care. Tesla reported "exploding" bottles in his salt-water capacitors. I have not experienced this in mine, but take precautions, especially if you stress the capacitor with exceptionally high voltages.

High-voltage capacitors can hold their charge for long periods even after the device is turned off and thus present a shock hazard. Discharge the capacitor of a Tesla coil by shorting out the spark gap. I use a screwdriver tip while holding the insulated handle. I see the telling spark with my commercial capacitors but not with my comparatively leaky homemade.

spark gap

The simple gap in its elementary two-electrode form will give you a Tesla coil that works, but any improvement you can make in this component will result in a boost in output and a stabilizing of frequency. The spark gap is really a type of semi-conductor. Ideally, it conducts suddenly and returns to nonconductivity immediately so the capacitor can charge right up again for the next firing. However, the air between the discharging electrodes becomes heated and offers a comparatively low resistance path for the current. This results in an arc being formed which prevents the capacitor from properly recharging.

To upgrade performance you must quench that arc. Tesla gave a huge amount of attention to perfecting the spark gap, which he regarded as a necessary evil.

A friend, Jim Campos, who is an electronics inventor, took an interest in the Tesla coil project and invited me to set up my coil in garage space next to his lab. Jim and I experimented with

perfecting the gap. From Jim's oscilloscope, we ran a long wire to within ten feet of the coil as a sort of antenna, for we did not have a high-voltage probe and it would not have been wise to connect seventeen hundred dollars worth of sensitive electronics directly to the output of a Tesla coil.

We fired up the Tesla coil with its noisy two-electrode gap, and the scope traces were revealing. Depending on the gap adjustment, the coil was putting out frequencies of 1/2 to 2 MHz, but these high frequencies only occurred in short bursts. Between the bursts the vibration dropped to zilch. The bursts themselves came at a rate as low as 2,000 per second. I had been getting shocks off the output, which I should not have, and this explained why. The high frequencies (above 2 or 3,000 cycles) won't register on the nervous system, but the low-frequency "envelope" packed a wallop. The problem, we surmised, was the gap, and indeed each little improvement we made there stabilized the frequency, increased the output, diminished the shock problem, and reduced the noise. The primary circuit, fouled by an arcing gap, was not pulsing the secondary in a rhythmic resonant fashion. It didn't swing. Imagine its performance with better quenching of the gap.

There are several ways to quench the gap. Any provision that takes away the heat of the gap, like cooling fins, helps to quench. Tesla found that directing a stream of compressed air into the gap blows out heated gases and helps to quench. Tesla also employed a magnetic "blow out" using a magnetic field to quench the arc. Magnets flank the gap, putting a North-South field across it. The magnets are stationed close to the gap and are shielded from it with little sheets of mica.

recipe simple spark gap

Parts for the recipe simple gap shown are available at any hardware store. The electrodes are 1-inch segments of 8/32" brass all-thread fitted with brass cap nuts. These don't hold up very well and have to be burnished after every 15 to 30 minutes of use depending on the energy levels handled.

A versatile low-power simple gap can be contrived from a set of automotive ignition points. This is my gap of choice for the oil coil, which is used primarily for electrotherapy. The tungsten points are very durable and go on and on with only an occasional cleaning with a miniflat file.

Tesla's patents show eight high-speed rotary "circuit controllers," all were mercury-actuated.

series gap

"The dielectric strength of a given width of airspace is greater when a great many small air gaps are used instead of one," said Tesla. The principle evolved into the series gaps that were commercially manufactured for spark radio transmitters. An engineering rule of thumb: a gap should be under 1/100th of an inch, 1000 to 1200 volts per gap. Breaking down the simple gap into a series of in-line pairs of electrodes greatly contributes to quenching. A good series gap can be improvised from spark plugs. Use only non-resistor plugs (not so easy to find these days). The gap shown is built on polycarbonate plastic.

Though a spark-plug series gap does not run silently or ozone-free, it is my present gap of choice for all-around simplicity, availability of parts, and durability. Use non-resistor plugs. I use the narrow AC 44TS. These require a hole of 13 mm (.5118 inch), and you can force-thread them into the plastic block. Since special drills are needed for plastics, I had the plastics supply shop drill eight 1/2-inch holes and widened them out myself with grinder and file. Acrylic will do for the block, but I used polycarbonate because it can take 90° more of heat. You can gap the plugs precisely with a feeler gauge. They will need to be regapped and cleaned periodically.

An evolved form is the air-tight spark-plug series gap. The project shown is made of ultra-high-molecular-weight (UHMW) plastic. The 1/2-inch head is drilled so that the plugs can self-thread into the UHMW. The block is drilled into cylinders of larger diameter. The head gasket is a punched-out sheet of high-temperature red silicone.

The air-tight gap runs almost silently. The plugs last longer because there is only a small amount of air within the cylinders to convert to corrosive ozone. The gaps can be made oxygen-free if you can contrive to close up the assembly in an exhausted atmosphere or in a chamber of inert gas.

I experimented with an airtight series gap in which the electrodes were made of brass end caps. The insulators were from the technology of the potter.

The one-inch ceramic post is an item of kiln "furniture" suggested by a potter. It's called a one-inch post and has a hole through it that can accommodate the electrodes. These posts are used to prop up pots in kilns, and you can get them at a ceramics supply store. The gaskets are cut from another ceramics item, a refractory called A-970 paper, and we also tried a refractory cloth called Fibrafax. This gap runs silently and ozone-free, and its

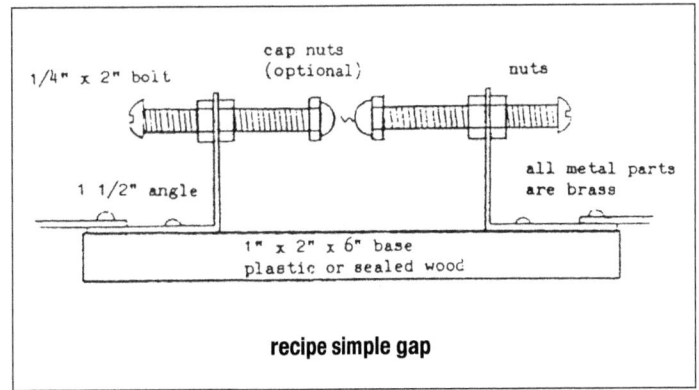

recipe simple gap

auto-ignition-points adjustable gap

spark-plug series gap

quenching effect is visible in increased output. Its limitation is again in the brass electrodes and their corrosion.

Although rotary gaps are often preferred by builders of big Tesla coils, the series gap is simpler, easier to build with precision,

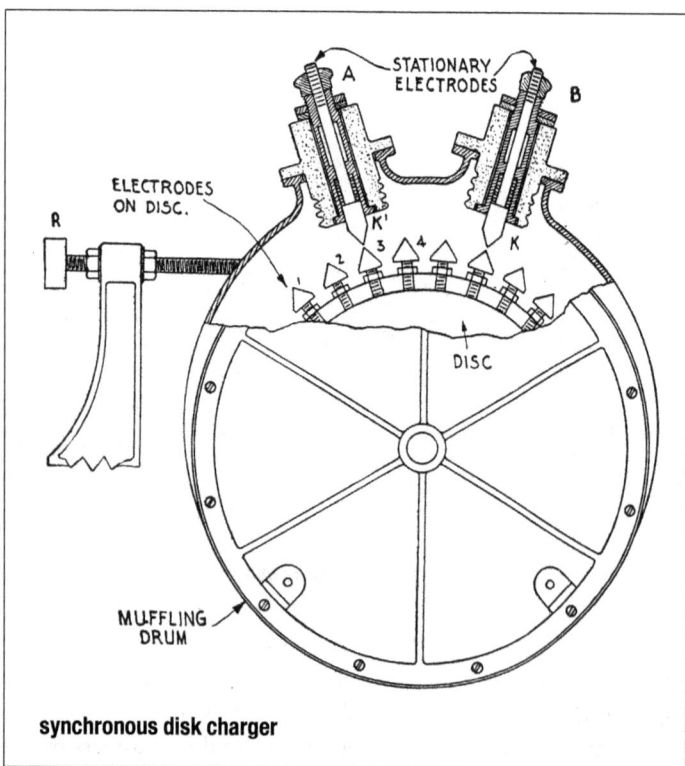

experimental series gap

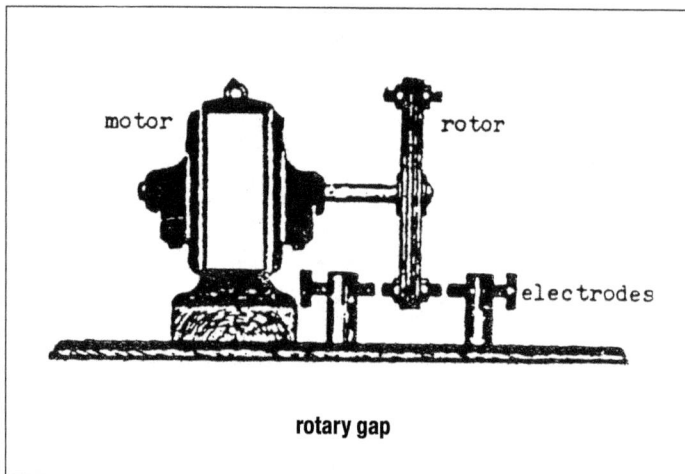

rotary gap

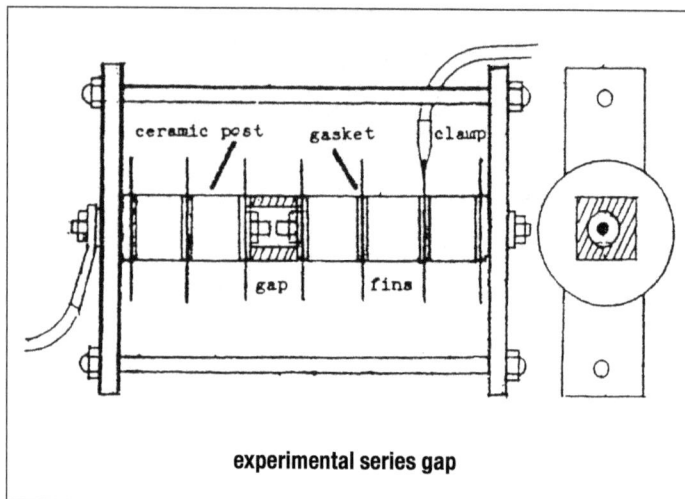

synchronous disk charger

and there are no moving parts. You probably should not attempt to build a rotary gap unless you have some machinist skills and equipment.

The current-carrying capacity of the series gap can be considerable, as is demonstrated by early high-frequency induction-heating spark-gap oscillators or "converters," as they were called. Industrial induction heaters were rated up to 15 kw and could have as many as 30 spark gaps in series — gaps consisting of one-inch tungsten disks, water-cooled. The spark-gap converter of an induction-heating unit is a simple circuit that resembles the primary circuit of a Tesla coil. These were replaced by vacuum-tube oscillators, which resemble tube-type Tesla coils. Get a hold of either type and you could drive a Tesla coil to the moon.

rotary gap

The rotary gap has spinning electrodes that break the circuit before arcing can take place. Tesla patented a number of sophisticated rotary gaps, including rotors immersed in flowing oil, rotors that dip into pools of mercury, and even mercury jets, but a good rotary gap can consist of brass studs on a plastic disk that is spun between two adjustable electrodes (like those used in the simple gap) by a small electric motor. Use a sturdy disk at least 1/4-inch thick and about six inches in diameter. On the perimeter mount 20 studs cut from brass all-thread, the thicker the better, and held on with brass nuts. Again, a tougher metal than brass would be superior. I have heard of carbide electrodes being used in such a gap.

A ready-made rotary gap may exist in the auto distributor, which is pure Tesla. In fact the old automotive ignition system is very Tesla.

The speed of the motor driving a rotary gap can be varied. This gives greater control over the rhythm of the primary circuit.

author's air-tight spark-plug series gap.

A variac is ideal for an AC motor , but you may be able to achieve some control on the cheap with a light-dimmer unit.

As with the series gap, the most advanced rotary gaps were invented for the spark radio transmitter. The synchronous discharger shown was manufactured by the Marconi Company (drawing from Bucher, *Practical Wireless Telegraphy*, 1917).

tube-type coils

The Tesla coil has evolved little since Tesla's time, but, not surprisingly, one significant development eliminates the problematic spark gap altogether. This development is the vacuum-tube Tesla coil, which uses a conventional oscillator circuit to pulse the primary coil. I have never seen one of these in action, but they must deliver smooth-running performance compared to the spark-gap type. They would also seem to be far more adaptable to modulation for voice radio transmission.

In fact, the tubes these Tesla coils use are radio transmitter tubes, big triodes. To get sufficient power for larger coils, two or more are ganged in parallel. To run these tubes, you need a step-up transformer to supply a potential to the tubes' plates of 800 to 3000 volts, and you need a filiment supply of 5 or 10 volts. Some of the tubes I've seen specified in various projects: 10Y, 100TH, 21 1, 304TL, 304TH, 801A, 807, 810, 811A, 826, 833A, VT-4C, and 3-5002. One source: 807's are sold by mail from Antique Electronic Supply, 688 W. 1st St., Tempe, AZ 85281. Price: a reasonable $7. A reader told me he got an 811A on special order through Radio Shack for $22. An old tube manual in the library will give the voltage info, but remember that you can drive the plate about 30% over the rated voltage. Look for transformers in electronics surplus stores. The value of the capacitor in the plate circuit is varied in the tuning process.

If the spark gap can be defeated by the vacuum tube, can it also be defeated with solid state? The answer is yes. MOSFETs are available that can handle enough power to drive small demonstration coils or to help tune larger coils at low power. Heavy heat sinking is required for these high current levels.

the terminal capacitor

The terminal capacitor is that ball or whatever that sits on top of Tesla coil secondaries. Metallic and usually hollow, the terminal capacitor is a sort of antenna that draws up and radiates the energy of the secondary coil. In fact, if the Tesla coil is to be used as a transmitter, the terminal capacitor is the antenna, only in this case it is set high above the ground as an "aerial capacity," as Tesla called it. On his Colorado Springs magnifying transmitter, Tesla used a 30-inch ball made of copper foil over a wooden form. It was coated with an insulating layer of rubber. The capacity of the terminal is a factor in performance and should be experimented with. A rule of thumb is that the terminal's diameter should be as large, if not larger, than the diameter of the secondary coil. Used for terminals are: hollow brass door knobs, polished world globes, pairs of stainless steel salad bowls, and I've read of one experimenter who made a toroid (donut-shaped) terminal by connecting together four chimney pipe elbows.

recipe terminal

Yes, what you see on top of the recipe coil is a copper toilet float. A 1/4 x 2-inch brass machine screw fits into it nicely and will hold it on to what you use to plug the top end of the secondary. I used a plastic item from the hardware store's plumbing shelves called a 3-inch plastic ABS test cap.

the recipe circuit

The complete Tesla coil circuit, including chokes, safety gap, and on-off switch is shown on page 86. The following should be connected to a hefty ground terminal on the chassis: the bottom lead from the secondary coil, the center electrode from the safety gap, the transformer case, and the green ground wire from the three-wire power cord. Unless you know for sure that the green wire leads promptly to a good earth ground in the house wiring system, you should attach a heavy conductor to this terminal and run it to a cold water pipe or to a conductive rod sunk three or more feet into the earth. Soak the earth around the pipe or rod with water to promote better grounding. Good grounding is important for safety, transformer protection, suppression of TV and radio interference, and performance. Tesla's ground at Colorado Springs consisted of a 20 x 20-inch copper plate sunk twelve feet into the earth. Over the top of the plate he spread a layer of coke. He ran water over the spot continuously. Any body of water makes a good ground.

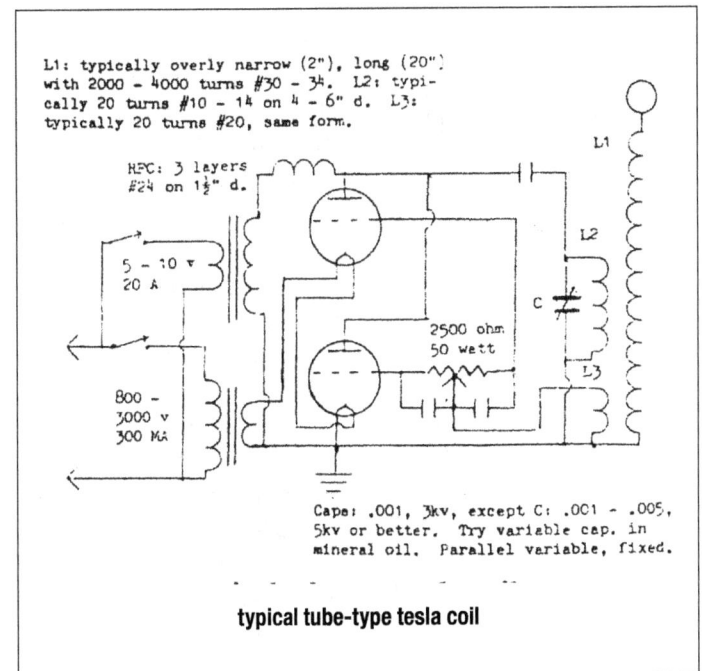

L1: typically overly narrow (2"), long (20") with 2000 - 4000 turns #30 - 34. L2: typically 20 turns #10 - 14 on 4 - 6" d. L3: typically 20 turns #20, same form.

RFC: 3 layers #24 on 1½" d.

5 - 10 v
20 A

800 - 3,000 v
300 MA

2500 ohm
50 watt

Caps: .001, 3kv, except C: .001 - .005, 5kv or better. Try variable cap. in mineral oil. Parallel variable, fixed.

typical tube-type tesla coil

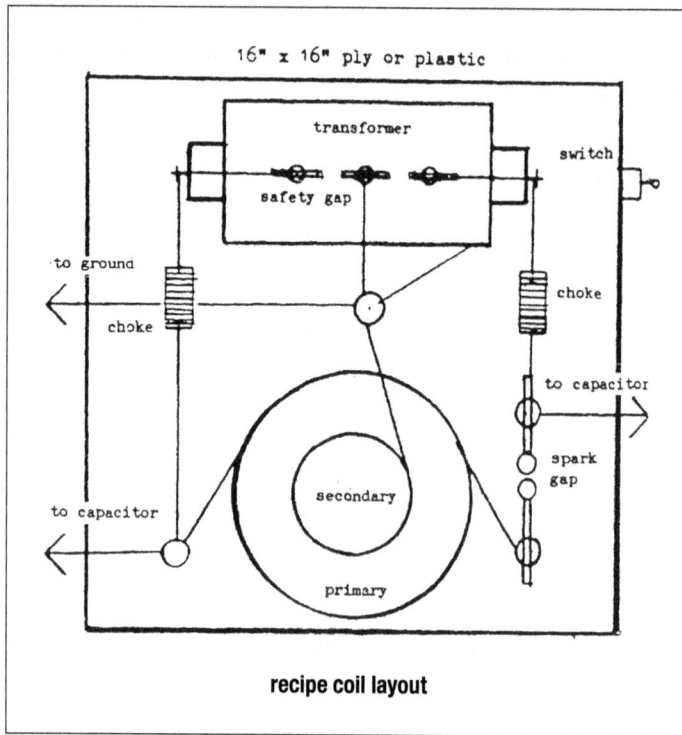

recipe coil layout

construction notes

For a structure on which to mount the components, use plastic sheet or plywood, dried and sealed. You can simply mount all parts on a single level, breadboard-style, or build a two-level chassis as I did for the recipe coil. Keep all wire connections as short as possible but keep the components separated enough to prevent arcing. For a good hook-up wire try neon cable; it's rated 15kv.

The two-level recipe chassis is constructed of 1/4-inch plywood on a frame of I by 2's. I put a sheet of plastic on the underside of the top level to ensure insulation between capacitor bottle terminals and components on top, but this may not be necessary, particularly if there is sufficient vertical clearance. The transformer is mounted with heavy (5/8-inch) bolts, the rear set of bolts passing through the frame. Under the secondary coil, I glued on a piece of 3/4-inch board to provide a good foundation for coil mounting. The coils

turns per inch table

wire#	enamel	insul.*
8	7.6	7
10	9.6	9
12	12	11
14	15	14
16	18.9	17
18	23.6	20
20	29.4	24
22	37	30
24	46.3	36
26	58	44
28	72.7	52
30	90.5	60
32	113	68
34	143	78
36	175	90
38	224	100
40	282	111
approximations Insulation thickness varies.		

are attached by means of a turned plug that fits snugly into the bottom of the secondary form and pegs through holes drilled in the bottom of the Serve 'n Saver into the chassis. You can make do with any wooden or plastic thing that can be jammed into the bottom of the secondary form so it can be attached. If you use wood screws or bolts, make them brass. It is a good idea to mount hefty rubber feet under the chassis.

hazards

The entire primary circuit is a shock hazard and should not be touched while in operation. To safely adjust the spark gap while running, Jim contrived a 15-inch long handle of narrow plastic tubing with a screwdriver head embedded in one end. A good practice honored by electricians at work around hot circuits is to keep one hand in a pocket at all times.

The secondary terminal should put out harmless high frequencies but may carry a low-frequency component that could shock. When well tuned, you can hold a piece of metal in your hand while streamers play off of it, and streamers can arc painlessly to the bare skin. However, limit skin exposures to brief periods and don't stick your head near the terminals because your eyes could be injured by streamers.

That faint chlorine odor around a sparking coil is ozone (O3). This super-oxygenating gas will purify the surrounding air, but at high concentrations can irritate the bronchi and lungs (like smog). Run the coil in a well-ventilated area.

The spark gaps can give off strong ultra-violet light; avoid sustained exposure to eyes. Contrive a glass shield for the gaps if possible.

Energy from a potent coil can disturb or even dud at a distance of many feet sensitive solid-state circuitry. Tesla coils and computers don't mix. It would probably be a good idea to disconnect your PC from the power socket while operating a Tesla coil and to keep floppy disks and other memory units away from the Tesla coil's ambient energy.

Tesla coils can adversely affect the functioning of a cardiac pacemaker. Announce this information before demonstrating a large coil before a group. If you wear a pacemaker, try another hobby.

Capacitors are a shock hazard. See capacitor hazards above. Discharge capacitors before touching.

You may put a neighbor in a hazardous mood because of TV or radio interference. Most interference goes out over the power lines and can be prevented by putting a brute-force line filter between the wall plug and Tesla coil. This will also help to suppress any kickback of high voltage into the power lines that might make it past the grounded safety gap.

If neighborhood complaints persist, the remaining "ambient" interference can be dealt with, to some degree, by shielding. The entire device can be enclosed in metal screen (Faraday cage) or in a sheet-metal container which is grounded. High-voltage lines running from the Tesla coil to experimental lighting fixtures, etc., should ideally be shielded and the shielding grounded. However, shielded cable for such high-voltage applications is not easily found and may have to be improvised. Incidentally, shielded cable is a Tesla invention.

tune the coil

A gauge of performance is streamer length and intensity. Connect a length of stiff heavy wire to the ground terminal, bend it so it curves around a safe distance from any of the apparatus and ends up pointing at the secondary terminal from about six inches away. Strip about an inch of insulation off this end and file a point on it. Turn on the coil and watch for length, regularity, and brilliance of the streamer as you adjust the spark gap, put in more capacitance, move around the adjustable primary tap, if you have one, and experiment with terminal capacitors.

As you fiddle and tune, you can hear the parts come into harmony; the machine hums. For most adjustments, you'll have to turn the coil off and discharge the capacitors to avoid shock hazard. In an 1896 patent, Tesla shows a way of fine-tuning that I have not seen implemented by builders, although the principle is basic to radio tuning. Across the secondary coil he puts

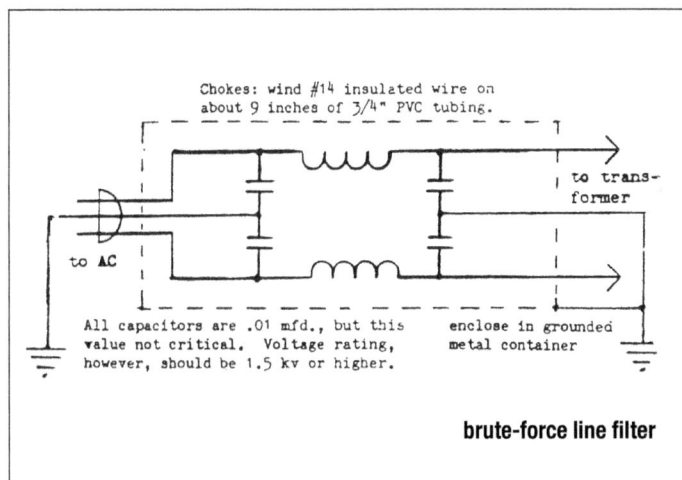

Chokes: wind #14 insulated wire on about 9 inches of 3/4" PVC tubing.

to AC

to transformer

All capacitors are .01 mfd., but this value not critical. Voltage rating, however, should be 1.5 kv or higher.

enclose in grounded metal container

brute-force line filter

a variable capacitor consisting of two cymbal-like disks that face one another and are adjustable as to distance apart (see page 105). He gives no clues as to their dimensions, but notes that at these high potentials not much capacitance is needed to affect performance. The same patent shows tuning by means of a variable choke in the primary circuit. An iron core, movable within the coil, varies its inductance and the primary frequency.

CHAPTER 2

Build a 3rd Generation Tesla Coil

a third generation?

The oil coil is a "third generation" coil. You can find all three generations of Tesla coil in Tesla's own work, but one particular configuration, the secondary wound with a single layer of fine wire on an elongated cylinder, has dominated Tesla coil building by experimenters for nearly a century. Why? Perhaps the inspiration came from the publication of the sensational *Century Magazine* photos of 1900 showing Tesla's Colorado Springs helical coils. This image, in the absence of others that might have been published had Tesla's work not been otherwise suppressed, may have dominated the public: consciousness to the exclusion of other designs that might have caught on.

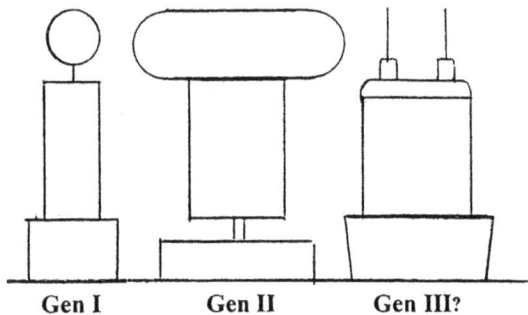

Gen I   Gen II   Gen III?

a Tesla oil coil

author's Gen 2 coil

Generation I

We built it in the previous chapter. A long, narrow helical secondary coil; a helical primary of several turns of heavy wire set far apart from the secondary (loosely coupled); the terminal, a sphere or whatever, relatively small in diameter: with some notable exceptions, this basic format dominated Tesla coil building through much of the 20th Century.

Generation II

The Gen II coil evolved the 1980s and is now the dominant fashion among the cognoscenti of Tesla-coiling. This evolution retains the traditional single-layer helical secondary, but now there is a greater appreciation of the larger diameter secondary in proportion to length. The primary is likely to be a flat spiral but may be a cone-like helical, expanding outward as it rises.

A close coupling between primary and secondary is appreciated more than ever, and, if it can be mechanically devised, the separation between primary and secondary is changeable, a variable coupling. The Gen II coil shown here, however; is a miniature (secondary only 8 inches long by 3-inch diameter), and with the toroid is resonant up at 540 kc, so it has never seen one-to-one pulsing on a spark system.

As Tesla coil building evolves, there is greater appreciation of matching of the fundamental frequency of the primary pulsing system with the resonant frequency of the secondary. This is way more manageable with solid-state pulsing systems than with spark-gap systems, as we shall see.

Generation III

The Tesla coil I'm calling Gen III is inspired by the coil demonstrated by Nikola Tesla at his London Lecture of 1892, and, who knows, if this lecture and the drawings published with it (in 1894 by James Commerford Martin) had been more widely circulated through subsequent decades, then this model, through more difficult to construct, may have become the experimenter's standard.

The proposed Gen III Tesla coil is a radical departure. Its format has more in common with the old induction coil than with any of the traditional single-layer helicals. It is a multilayer coil. The secondary is wound right on top of the primary, closely coupled, not loosely.

How can this be? My solid-state mentor, Jim Campos (who has worked with me on spark systems as well as transistor, and from whom we will hear much more later) argues that loose coupling is only necessary to compensate for the vagaries of spark. If the primary circuit goes completely off then it is "isolated." There is no damping of the secondary oscillation.

From the standpoint of efficient transfer of energy, of course, a close coupling is desirable. A close coupling is especially desirable when pulsing the primary coil with relatively low voltages, as will be the case when pulsed from solid state circuitry.

oil immersion

With a multilayer secondary and close coupling, insulation becomes critical. The solution is immersion in oil. Says Tesla, "As sparks would soon destroy the insulation, it is necessary to prevent them. This is best done by immersing the coil in a good liquid insulator, such as boiled-out oil. Immersion in a liquid may be considered almost an absolute necessity for the continued and successful working of such a coil."

Liquid insulation is self-repairing. This self-repairing property can be obtained to a degree even with the conventional open-air helical if you spray on heavy layers of silicone. Spray silicone can be found in the lubricant section of your hardware store. Oil immersion is feasible for a Gen II helical secondary providing that a second concentric tube can be slipped over it, sealed, and oil poured in between. I have that plan for my little

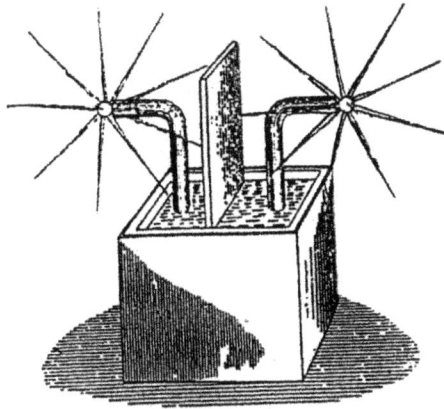

Tesla's London coil

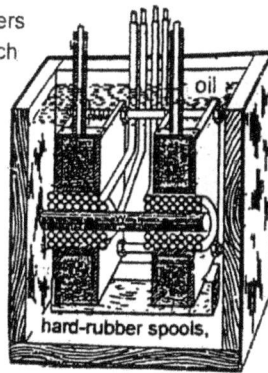

sec: 26 layers
10 turns each

two sec's wound in opposite directions

pri: 4 layers
24 turns each

hard-rubber spools,

Tesla's London coil, detail

close coupling

Gen II helical shown, although the variable-coupling feature makes sealing in the oil an engineering challenge.

Before a spell-bound audience of electrical engineers in London in 1892, Tesla drew from his oil coil streamers of fire and strange white luminous brushes. "A most curious form of discharge is observed with such a coil," he noted.

In praise of oil, Tesla said, "I am led to believe that in our future distribution of electrical energy by currents of very high tension, liquid insulation will be used." Tesla envisioned a high-voltage coaxial cable which contained an oil insulation. "The cost is a great drawback," he continues in his lecture, "but if we employ an oil as insulator the distribution of electrical energy with something like 100,000 volts, and even more, becomes at least with higher frequencies, so easy that it could hardly be called an engineering feat (even) at distances of as much as a thousand miles."

Experience with my own oil coil suggests that an empowerment is obtained exceeding that obtained by the mere insulating effect of oil. There is a dividend, a special advantage, secured by oil's dielectric dynamics.

Is the Tesla coil really a capacitor?

Conventional electrophysics construes the Tesla coil in electromagnetic terms, the primary and secondary as inductances. And, oh yes, a whole mathematics is worked out accordingly. But Tesla researcher Gerry Vassalatos, who has studied some of Tesla's unpublished notes, says Tesla early on experimented with "coils" that were capacitive, being smooth copper cylinders. These he later spiral-grooved on a lathe, increasing greatly the area and hence capacity. Next came bare-wire windings, which had a similar and greater effect. Did the helical thus evolve?

The dielectric constant of oil is more than twice that of the driest air, the puncture voltage ten times that of air. In his

lecture Tesla also recommends for his oil coil a variable condenser (capacitor), which "should be an oil condenser by all means, as considerable energy might be wasted." In his patents Tesla shows the entire system, including the spark gap, in flowing oil.

My oil coil, which is a simplification of Tesla's London oil coil produces the whitest, hottest, loudest arcs of any I have built. I have become a believer in oil magic.

coils that do real work

Did Tesla invent his coil just to throw sparks? One might think so given the direction Tesla-coil experimenting has taken over the decades. No doubt Tesla enjoyed his coronas, sparks, and streamers as much as we do, but Tesla's effort was directed at the development of high-efficiency lighting plants and other practical equipment in which arcs and coronas could be destructive forces. This will be our effort, too.

Engineering for pyrotechnics teaches us some electrical-engineering lessons on the controlling of awesome potentials, but, in designing systems that can do practical tasks, we honor some important engineering criteria that in designing to throw sparks we customarily throw to the winds. These are:

1. low current draw
2. long duty cycle
3. low maintenance
4. quiet operation
5. emission-free operation
6. compactness
7. safety

These criteria amount to a big challenge to the builder. In disruptive-discharge systems, the spark gap alone presents some tremendous engineering problems. Tesla invented the resonant transformer that came to be called the Tesla coil for the specific

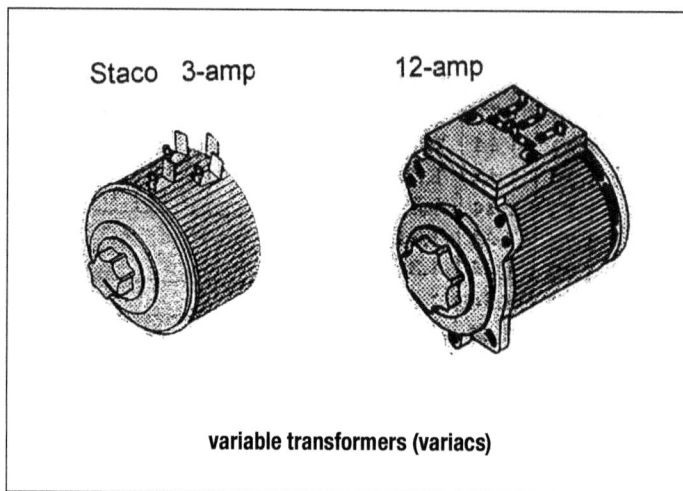

Staco 3-amp 12-amp

variable transformers (variacs)

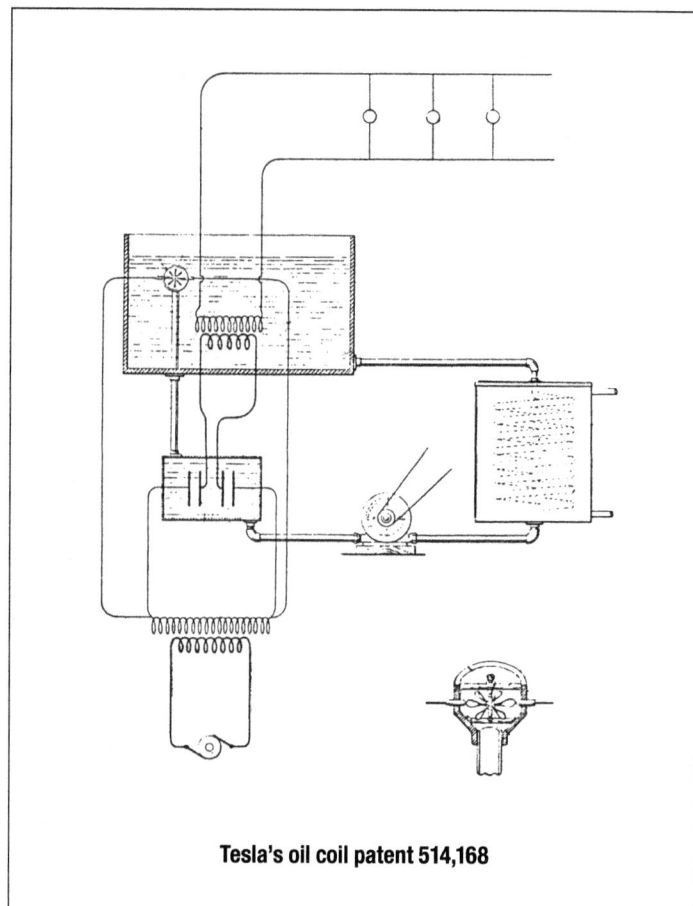

Tesla's oil coil patent 514,168

purpose of producing high-voltage, high-frequency currents to stimulate with unprecedented efficiencies appliances that would produce light. Let's pick up where that tradition leaves off. With an eye to the above seven criteria, we will construct a high efficiency oil-coil lighting appliance that can run from batteries.

Below you will learn how to build a solid-state pulse generator for a Tesla lighting plant that can drive fluorescent tubes from DC Sources 12 volts and up.

in praise of coils more tame

It is amazing what can be done with relatively low power Tesla coils. It is the tamer coil that we can put to practical tasks. Low power Tesla coils also provide some unexpected benefits for the experimenter. Try a variable transformer between the wall socket and that big supply transformer on your monster spark-thrower. Tone down that noisy, scary machine to a quieter more modest level and you might discover a whole new world of experimental delights. Experimenters often refer to the variable transformer as a "variac." Variac is a brand, as are Powerstat, Superior, Ohmite, and Staco (shown).

Have you observed the phenomenon of the sparkless spark gap? Neither had I until I tamed it down with a variac. High-frequency electrotherapy (Appendix A) is done with a Tesla coil of restrained output . Many adventurous experiments can be conducted quietly and safely. For low-power experimenting, I built what I call my "dream' power supply around a small Powerstat variable transformer (270 VA), a small open-core neon transformer (5 kv, 20 MA), and a bridge rectifier. The outputs are 0-140 volts AC, 0-5,000 volts AC, and 0-90 volts filtered DC.

Practical Tesla coils are smaller and tamer than what many experimenters have become accustomed to. All over the world there are big macho machines sitting (mostly idle) in basements and garages, and these monsters by themselves can generate enough Tesla power to light up the neighborhood. But, for local purposes in the home, farm, or small workplace, what is appropriate is the modestly scaled coil.

build an oil coil

If you still want to, build a veritable flame-thrower, if you are an unrepentant builder of macho coils where value is measured in coronal and streamer pyrotechnics, read the following but ignore my recipe and instead challenge yourself by building Tesla's 1892 London oil. My recipe oil coil is simpler and easier to construct. It is designed for failure-free constant running in the spark mode with only 5kv input from a transformer, and, in the solid-state mode, will probably never see much over 50 kv at the final output terminal. The recipe oil coil is resonant at 460 kc unloaded and, loaded with a four-foot florescent tube, at 115 kc.

Since both Tesla's original and my recipe coil are oil-insulated, what is the difference that enables Tesla's to produce grander voltages? Tesla's original has more layers, and the width is narrower. This engineering respects the hazard of inter-layer arcing. The narrowness allows less voltage increase per layer and hence less voltage difference between layers. If you are building for pyrotechnics, you should also imitate the best you can the thick insulation Tesla recommends in primary and secondary conductors. If you go all the way with Tesla's original and provide two coils, note that they are wound in opposite directions. Evidently, this provides for push-pull action. If you succeed with this coil, it should be dynamite. Please send me photos.

author's 5 KV power supply

author's oil-coil winding jig

preparations

Whichever oil coil you make, the recipe or the true London, the insulation of off-the-shelf wire will be inadequate layer-to-layer, so for the oil coil you will wind on between each layer the material Tesla recommends, which is oil-soaked muslin. Muslin means cotton, as in bed sheets, so you will need to scrounge an old bed sheet or, in desperation, a new one (if she's not looking) and rip the thing, in the direction that it wants to rip evenly, into strips of about two inches in width. Make yourself a stash of these in lengths of about ten inches or so.

The oil can be automotive oil; that's the cheapest liquid insulation you'll find. Tesla used boiled-out linseed oil. I considered using one of the new synthetic motor oils since these are supposed to take more heat. That's a consideration since a tiny trapped bubble can heat up to 4000 degrees F. (Tesla notes, though, that such bubbles will be driven off by high-frequency currents. That is, if they are free to move.) Synthetics being expensive, I settled for a blend of conventional and synthetic, which was Quaker State 4x4. Regular motor oil (nondetergent) is cheaper and should work just fine. Get three or four quarts.

My oil coil is wound on a homemade spool the core of which is a 7-inch plastic pipe 1 ½ inches in outside diameter. The spool-ends, which you can improvise any way you want (they need not be perfect circles), I cut out of 3/16-inch ply using a circle cutter (fully extended) on a drill press. With the same cutting tool, I made a 1 ½ inch hole in the center into which I fitted the core pipe.

The job is best done on a winding jig like the one shown; in fact, it's hard to imagine doing the job without such a tool. The ideal way to wind an oil coil, Tesla's way, is with the whole coil and winding apparatus submerged in oil, thus assuring the complete displacement of air. You may have the capability of contriving such an arrangement, I didn't, but there are other ways of removing air. The process of winding on the wire is itself air-removing.

Needless to say, even unsubmerged, winding an oil coil can be a messy job. Put a tray under the coil to contain the profusions

of oil that will be dripping off as you wind. Put newspaper everywhere. Keep a roll of paper towels handy and a big metal wastebasket. By all means get yourself a box of latex surgical gloves (a drugstore item) and be prepared to change them frequently. Promptly dispose of all oil-soaked material in a safe manner that will not be prone to spontaneous combustion.

winding and wrapping

Put some oil in the tray beneath the coil and in it lay several of your bed-sheet strips. Wind some oil soaked strips onto the bare spool core and onto this wind the first layer of wire, the primary coil. My primary consists of just one layer of #18 very well insulated high-voltage 5kv test wire, very flexible and easily wound. (I have found many uses for this wire in high-voltage work.)

In my next coil, I may try multiple primary layers, and maybe, in respect to Tesla's principle of having primary and secondary of equal weights of copper, use a heavier conductor. A design alternative could have the primary as a separate coil which is inserted into the hollow, oil-filled core.

Drill the spool end and pass the wire through and wind on a primary layer or two. On top of the primary winding, wrap two or three thicknesses of your oil-soaked muslin strips. The breakdown between primary and secondary is a real concern, so especially if your primary wire is weakly insulated, lay on the oiled strips.

As you wind on the wire, make sure the coil is constantly saturated with oil so that it is wet and dripping. Baste the whole length of the coil while winding by dipping your latex-gloved fingers into the oil tray and dripping the oil onto wire and muslin. When you rest between layers, wrap the coil loosely in plastic sheet to keep dirt out.

My secondary conductor is plastic insulated, #20 solid, a fat wire as Tesla secondaries go but defensible since it's tough and resistant to breaking as it is stretched taut in the act of winding, and it is also capable of carrying lots of current. You may want

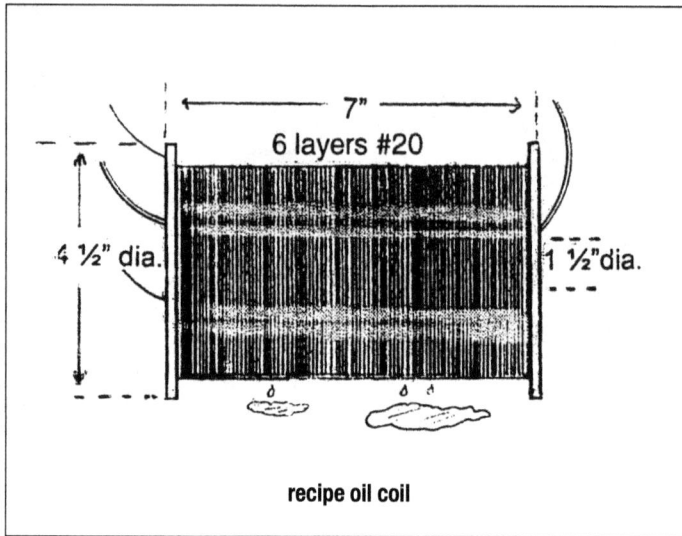

recipe oil coil

author's oil coil set up for spark

to use a smaller gauge wire, but a plastic insulation, rather than enamel, may be appropriate for a multi-layer coil. The #20 gives 15 turns per inch, 105 turns per, layer. Drill a little hole in the spool-end to let the wire in; you'll drill another at the end of the final layer to let the wire out. As you wind the wire on, keep it taut and it will squeeze out the oil as it progresses down the length of the form.

You'll be putting on six layers (total turns: about 630). Between each layer, you'll be putting on more oiled muslin, and, since voltage will build layer-to-layer, each successive secondary layer gets progressively more thickness of insulative cloth between. The final layer needs no cloth over it.

containment

Promptly remove the coil from the winding jig and slowly submerge it in oil. Use some appropriate insulative container. My first one was a plastic one-gallon pitcher from our friends at Rubbermaid. Later my container became a tiny aquarium of hexagonal cross section called a Minihex from Island Aquarium. (Tesla's container was a sealed wooden box with an outer lining of zinc sheet.) Run the wire leads out of the container to some appropriate primary and secondary connection terminals. **HAZARD NOTE:** Particularly if you are going to push this coil to the limit, and since your container may be of some fusible material, take the precaution of setting the whole thing in a metal container that has enough capacity to serve as a containment berm in the case of meltdown.

I have put into the primary system of the recipe oil coil as much as 12,000 volts at 30 MA for brief periods without incident. It is presently connected to a modest 5 KV, 20 MA transformer adequate for the electrotherapy function that it serves. I have yet to run this Tesla coil wide open to determine at what point, if any, I can make it fail.

CHAPTER 3

An Alternative Fluorescent

"**I** have made the discovery," Tesla announced in an 1891 patent, "that an electrical current of excessively small period and very high potential may be utilized economically and practicably to great advantage in the production of light." Tesla-coilers know that, under the influence of high-potential, high-frequency currents, a fluorescent tube in the vicinity just wants to light up; there is a natural compatibility. Hold a fluorescent tube near a Tesla coil terminal, and it lights up wirelessly. A fluorescent tube forgotten in some corner of the lab is sometimes unexpectedly seen flickering spontaneously, like it wants to get in on the fun.

You have no Tesla coil? You need only take a fluorescent tube into a darkened room to get a demonstration of its electrostatic receptivity. Just rub the tube briskly with a piece of fur or cloth, the hair on the back of your arm, your cat, or whatever, and it will produce little flashes of light. Fluorescent tubes want to light up under the slightest electrostatic disturbance, whether directly wired to a source or not.

conventional fluorescent

Compare the standard fluorescent fixture wired into 120-volt 60-cycle AC. The apparatus needed to light up the tube is surprisingly complex. By comparison, it's like the tube has to be clobbered into activity. A starting device is needed to provide a jolt at two to five times the line voltage, up to 600 volts, to awaken the tube. A clever little automatic thermal starting switch is required as well as step-up apparatus. The tube also requires a steady electric flow generated by the burning of hot cathodes. These erode slowly as they burn at 950 degrees F, and, of course, make the tube emit some heat. Like the filaments in an Edison incandescent, it is the on-and-off especially that fatigues conventional fluorescents. One start equals three hours of steady burning. One thousand starts kills them. The Tesla system suggested here has no use for these cathodes except as cold electrodes, so it's OK if the tubes are burnt out in the cathode department.

The starting jolt establishes an arc through the interior argon gas and mercury vapor, and the resulting ultraviolet light stimulates to luminescence the phosphors that coat the inside surface of the tube. (Caution, the phosphors within those fragile florescent tubes are toxic. If you cut yourself from the implosion of these fragile lamps, the wounds are particularly painful and persistent.) Full-spectrum fluorescents, for growing or for

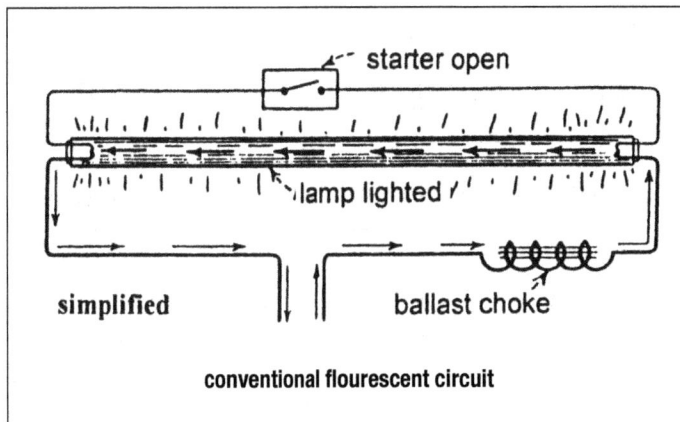

conventional flourescent circuit

environmental eye-ease, are painted with a mixture of different phosphorescent powders that produce a more natural sun-like emission.

ballasts

Once conducting, the tube has almost no resistance and would present a dead short to the circuit were it not for the so-called ballast transformer, which is not really a transformer in this function but a choke coil, and was called that in the early literature. The choke eats up some current itself, adding 10 to 20 percent to the wattage. A tube's rating does not include the ballast. For a tube rated 40 you can add about 8 more watts. The instant-start Slimline, has a ten-pound ballast that consumes ll watts. The ballast choke can also be used to magnetically store the starting jolt. When the starting switch opens, the field collapses, inducing a high-voltage surge. There may be a step-down transformer winding on the ballast used for a preheat circuit. Heavy ballasts require that florescent-tube fixtures be of heavy, cumbersome construction, although the tube itself is feather-weight and easy to support.

There is a small capacitor in the starting switch to suppress radio interference when the tube goes on and off, and there is another suppression capacitor across the line, and yet another may be used to put two lamps out of phase. You cannot dim these fluorescents. Special ones that dim require yet more ancillary hardware.

fluorescent sensitivities

Whether you drive them conventionally or Tesla-style, fluorescent tubes are sensitive to both temperature and humidity. An

Ventex 22kc 9kv 6.5"x1.9"x1 ¼"

electronic neon transformer

foil wrap
elec. neon transformer

electronic-transformer experiments

hot plazma
Tesla coil (6"x1"d.)
elec. neon xformer
h.v. diode
h.v. cap

Jim's torch

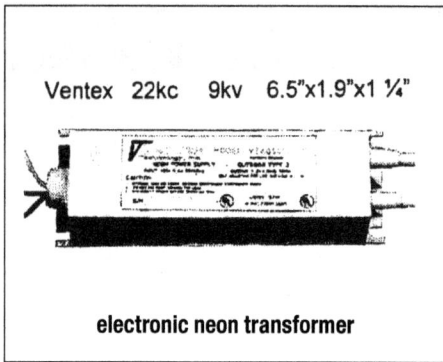

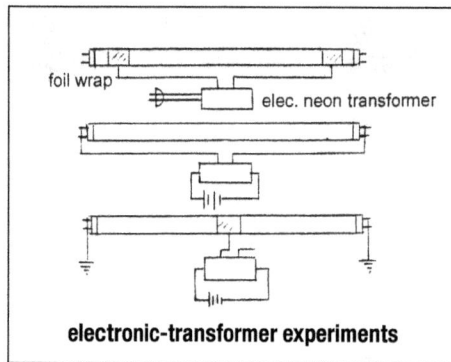

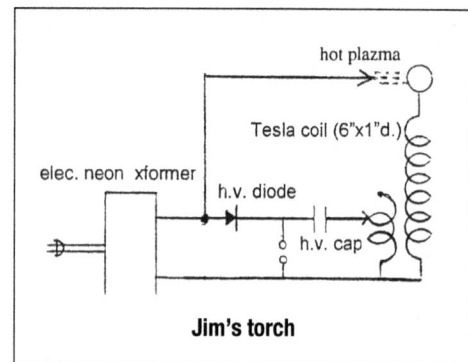

increase in humidity means more power is required to drive a tube, and humidity affects a tube's willingness to start up. In the conventional mode, the voltage necessary to start can vary from 250 to 750 volts, depending on humidity. Moisture on the surface degrades the dielectric quality of the glass so energy is not concentrated at the electrodes where it belongs. I have gotten brighter lights and easier start-ups in experiments with Jim in drier California than with tubes similarly driven in my own lab in rainy Oregon. To fight humidity fluorescent tubes are coated with silicone. This surface makes moisture ball up instead of forming a conductive sheen. Since this coating may be worn off of older tubes, I spray mine with silicone lubricant. Temperature affects the performance of these cranky lamps. A tube designed for indoor use will fail completely outside in the cold. Outdoor tubes wear a jacket of vinyl tubing. Such a sheath may also offer protection against humidity. Blackening results from mercury migration and streaking. Dimming can occur slowly over time due to phosphor decay.

How will the application of novel Tesla currents affect these idiosyncrasies? The higher voltages should facilitate starting, but humidity and temperature will still be factors. As to the affect on mercury streaking and phosphor decay, only usage over time will tell.

Fluorescent tubes even if driven by Tesla currents, are prone to flickerings and pulsings, often inexplicable, as are the standing wave patterns sometimes seen marching down the length of a tube.

the Tesla alternative

Early in the era of the Edison incandescent, Tesla said, "The more we progress in the study of electric and magnetic phenomena the more we become convinced that the present methods will be short-lived. For the production of light, at least, such heavy machinery would seem to be unnecessary. The energy required is very small, and if light can be obtained as efficiently as theoretically appears possible, the apparatus need have a very small output." Tesla was optimistic in believing that the old methods would be "short lived."

The contemporary compact fluorescent is stimulated by high-frequency pulsing. But pure Tesla lighting is both high frequency and high voltage, and, ideally, resonant as well.

electronic neon

High voltage is basic to neon. There is also a cold-cathode form of industrial fluorescent lighting that uses high voltage, up to 15,000 volts, the same as neon systems, and uses the same step-up transformers as in neon. Both neon and cold-cathode fluorescent systems are typically low frequency (60 cycles) and, of course, nonresonant.

The electronic neon transformer is the closest lighting has come to Tesla technology. These systems are high voltage and high frequency, running typically at 22 kilocycles and 3,000 to 5,000 volts. Most operate from a 120-volt AC source, but there are some 12-volt DC electronic neon transformers as well. My experiments with both the 120- and 12-volt versions suggest the possibility of using these for fluorescent-tube drivers. Can they serve as little off-the-shelf Tesla-style lighting plants? It may be worth pursuing.

Jim turned an electronic neon transformer into a hot arc-thrower that may have potential as a practical torch, perhaps a cutting tool. We burned out a couple of transformers doing stuff like this. The electronic neon transformer is a delicate item.

The high-frequency criterion of Tesla lighting is honored today in the compact fluorescent, which uses self-contained electronic pulsing circuits at high frequency (but low voltage) to produce a more efficient fluorescent. Off-the-grid home-power folks, who draw from solar-charged battery power with tight economies, can use 12-volt units.

Tesla invented the old 60-cycle system, and the Edison lamp became a fixture in it. Tesla reinvented his system, introducing patents for a system of high-potential high-frequency lighting for which the power supply became the Tesla coil. But, with the specialized exceptions above, the old system has become institutionalized, refuses to budge, and Tesla's high-frequency alternative has been cast out. Such social inertia does not alter the scientific reality that Tesla currents so effectively stimulate gas and phosphorescent lamps that the phenomenon may belong in another world of efficiencies. With the oil coil we will drive conventional off-the-shelf fluorescent tubes with currents that meet Tesla's criteria: high voltage, high-frequency, and resonant.

CHAPTER 4

Build a Solid-State Tesla-Coil Driver

spark vs. solid state

Experimenters defend the spark gap as the premier shock-excitation switch. No vacuum tube or semiconductor can match the ability of disruptive discharge to drive a Tesla-coil primary. Spark-gap systems can release great quantities of current with great suddenness, and the spark gap can withstand extreme electrical stresses. How would the spark gap have developed if it had not been abandoned by official electrical engineering circa 1920?

Still there is a case for solid state. The ratings of the commercial field-effect transistor, like the power FETs we will be using in the next chapter, are impressive for products of the relatively effete world of semiconductors: up to 500 volts and 25 amperes. Today there are integrated circuits and transistors that, in respect to rapid on-off switching, compete very well with spark. These have been invented and mass-produced to meet the needs of high-speed computing and are now inexpensively available for us to bend to our own perverse purposes. These new tools hold promise that an ideal solid-state Tesla-coil pulsing circuit can be designed from scratch.

the ideal pulse

My solid-state mentor, Silicon Valley inventor Jim Campos, pondered the question on his duel-trace scope. What kind of pulse does the secondary want from the primary to perform at its peak? A secondary coil has a particular resonant frequency. The primary ideally must match that vibration beat-for-beat. That's the ideal, but spark experimenters are accustomed to primaries that vibrate at s ome harmonic fraction of the resonant frequency.

The primary pulse should have an extremely sharp, almost vertical rise time. "Slam it like a brick," Jim said. Spark does fulfill that, but the pulse should not fall off in its own good time, as the spark pulse generally does, but be sustained for a particular period. Hence, pulse width, as well as frequency, must be precisely controllable over a wide range.

Looking at the primary pulse and the resulting performance of the secondary, the duel-trace scope told this story: The pulse must be prolonged enough to take advantage of the excursion phase in the secondary, the up-swing, but not so prolonged as to interfere with the down-swing. Such interference causes damping and disturbs the clear, steady ringing of a secondary in resonance.

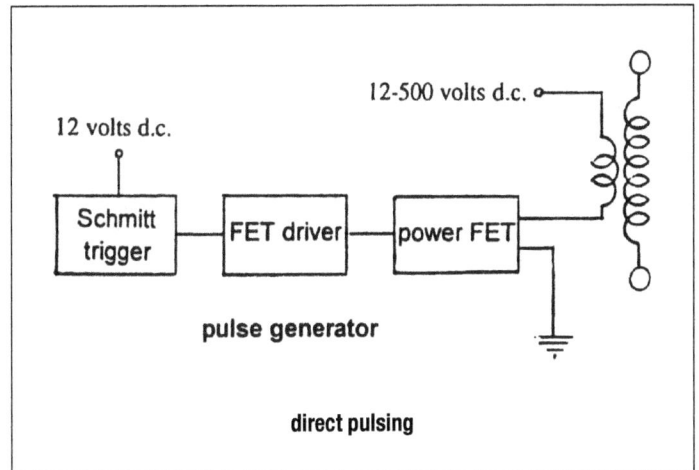

pulse generator

direct pulsing

If the primary pulses the secondary at exactly its resonant frequency beat-for-beat, the decay typical of intermittent pulsing is absent. The wave is not damped but is sustained as an undamped continuous vibration. The scope showed that the ideal period for sustaining the pulse is one quarter of the wavelength. For example, if the frequency is 250 kilocycles, the total wave duration is 4 microseconds, thus the pulse duration must be one microsecond.

Sharp rise, quarter-wave duration, sharp fall, a square wave. That's the ideal pulse.

solid-state chopping

A solid-state battery-current chopper seen among Tesla-coilers uses the 555-Timer chip in an astable mode to drive power transistors like the 3055. Both the 555 and the 3055 are widely available. The battery DC, chopped into a proper pulse, drives an auto ignition coil that serves as the high-voltage transformer for a small spark-gap Tesla coil. Mine had a chrome-plated hot-rod ignition coil. The solid-state driven ignition coil itself makes a nice laboratory induction coil or a high-voltage power supply for many experiments. The solid-state pulse circuit by itself makes a neat little audio tone generator. The little battery Tesla unit served as my medical coil for many years. The circuit shown was inspired by Jim, by Bob Beck, and by Walt Noon.

The 555 can be controlled for frequency by the turn of a pot, and the same goes for the control of pulse width. To drive an ignition coil, the frequency is set down in the audio range. (This circuit plus a loudspeaker makes a simple tone generator you can have some fun with.) The capacitor marked * in the illustration sets the frequency range, as do the values of the pots. Audio is in the lower range of the amazing 555, which can go up to one million cycles (or two million with the low-power 555). This ignition-coil system is a hybrid, a wedding of audio-frequency

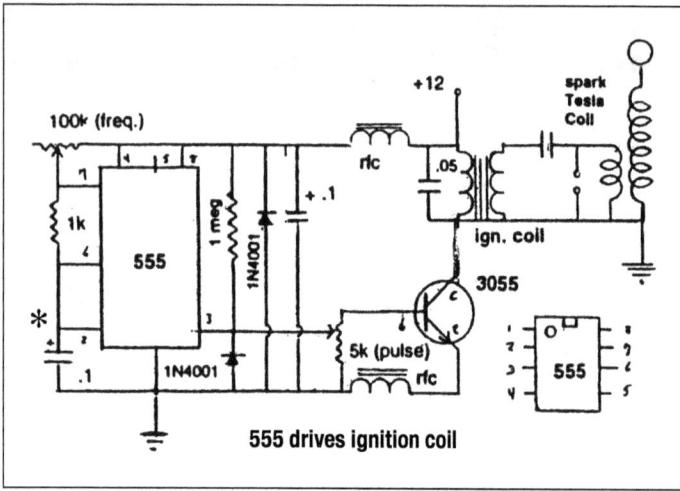

555 drives ignition coil

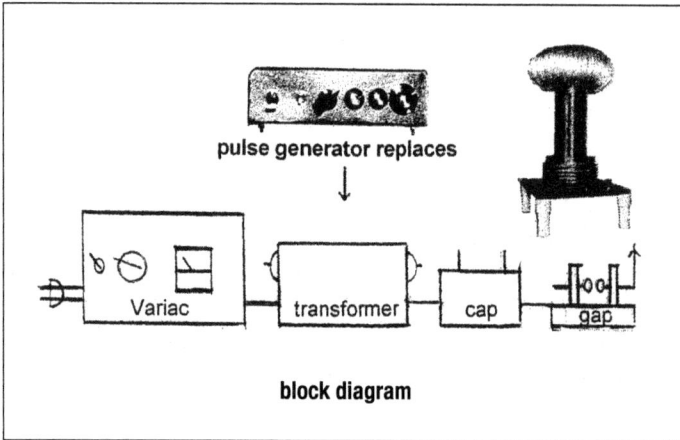

pulse generator replaces

block diagram

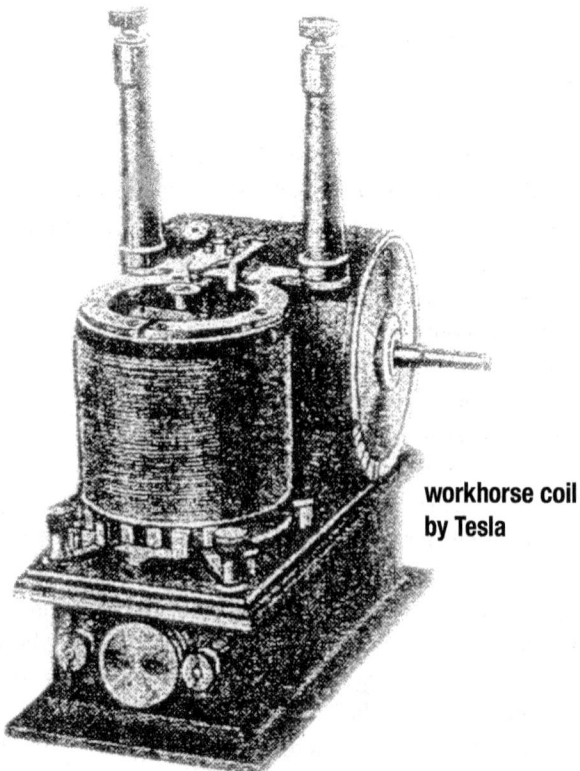

workhorse coil by Tesla

solid state and spark. The manufacture of faster transistors that can handle higher frequencies and of transistors that can handle higher power opened the way to a Tesla coil that could be pulsed directly by solid state.

recipe solid-state pulse generator

This section tells you how to build an experimental solid-state pulse generator that will drive directly, beat for beat, any closely coupled Tesla coil that is resonant at 50 to 300 kilocycles. The oil coil of Chapter 2 (when loaded) meets these specs just fine. The pulse generator weighs under half a pound but replaces the ponderous transformer, capacitors, spark gap, and variac of the conventional spark Tesla coil. It need not be plugged into the fragile grid and can run from batteries, if necessary, for it is powered by 12 to 48 volts of direct current.

Schmitt trigger or 555

The front end of such a Tesla-coil pulse unit can again be the old 555 workhorse. We tried the CMOS (the so-called low-power) version but it did not prove to be superior to the regular 555. We also experimented with the 4584 hex Schmitt trigger chip. The specs told Jim that these might be the superior slam-it-with-a-brick entries from the progressively more speedy slam-bang world of digital. Jim worked out all the details of this circuit in a series of experiments conducted over three years. Countless components and values were tried and rejected in oscilloscope drudgeries. An objective was to simplify and minimize the number of parts which now amount to about a dozen.

The 555 and 4584 can be wired into themselves to become high-frequency square-wave pulse-formers whose frequency and pulse width can be precisely controlled. The upper frequency limit of the 4584 according to the specs, is 2 megacycles. The 555, one MC. The 4429 chip is called a FET driver. The final output power transistor of choice is the IRF 640, rated 200 volts and 18 amperes, but other power FETs in the IRF series (and in other formats) could be used here if sufficiently speedy. The full designation for the hex Schmitt trigger is CD4584. Some suppliers know it only by its replacement-part number, which is NTE 23740. A similar package is the CD40106.

12 volts in?

It was not among our expectations that our experimental solid-state pulse generator would drive a Tesla coil with an input of under 50 volts or so. But we were delighted to discover that with a mere 12 volts of battery power we could get decent results. In fact, with just 12 volts input we could coax outputs from the Tesla-coil terminal of over 2000 volts. This was sufficient for lighting our four-foot fluorescent test lamp to practical levels. So here is clear testimony to the effectiveness of a closely coupled system pulsed one-for-one with the ideal waveform.

+12 V.D.C.

+12 to 25V.D.C. (The IRF640 will handle up to 200V.D.C. for when more exciting applications are desired.)

To lamps or other loads

Fuse 1 Amp.

S.P.S.T. SWITCH

Fuse 5 Amps.

5 Amp Meter

S.P.S.T. SWITCH 6.0A. MIN.

Secondary

1000uF. ELECT. 25V.

47uF. TANTALUM 25V.

Primary

Coarse freq. adj. 2K Trimpot

1K

2.7k

1N34

1N34

1K Dimmer

33uF TANTALUM CAP. 25V.

8 7 6 5

TC4429CPA

HIGH SPEED CMOS DRIVER

1 2 3 4

POWER MOSFET Harris IRF640 etc.

8 7 6 NC 5

NE555 TIMER

0.1 uF

1 2 3 4

1000pF

470pF

fine freq. adj. 2K

Pot Pin-out

Rear View

3 2 1

DRAIN FLANGE

SOURCE
DRAIN
GATE

IRF640

POWER MOSFET Pin-out

By:Jim Campos

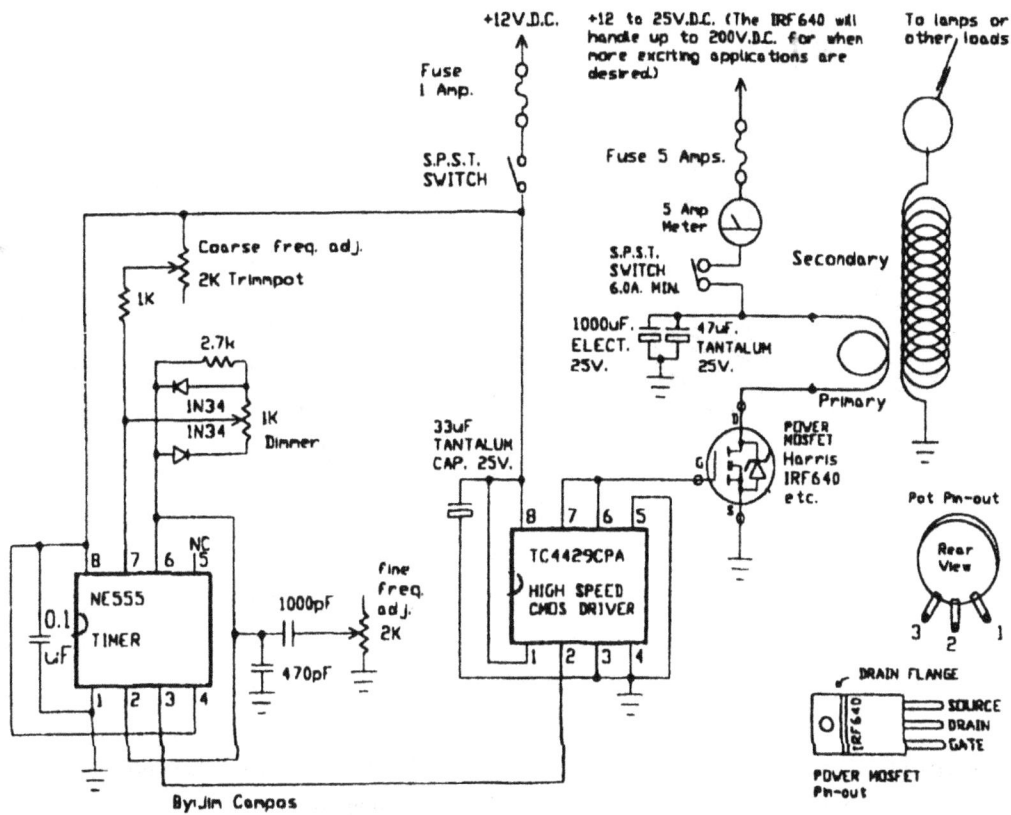

555 pulse generator

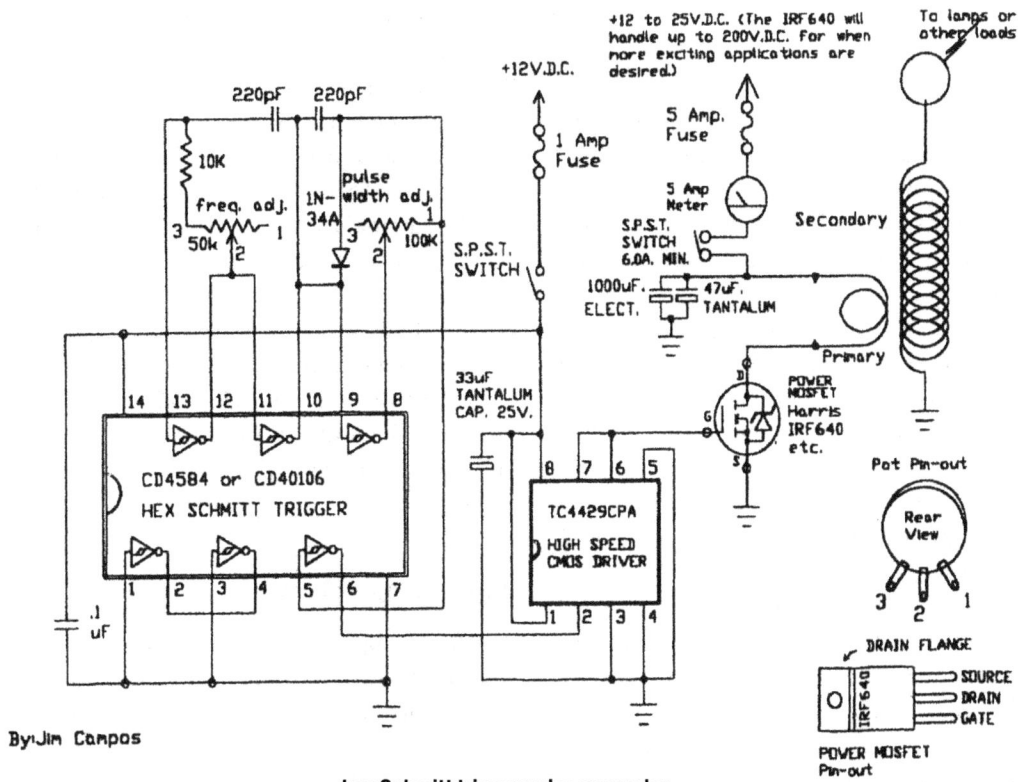

220pF   220pF

10K

freq. adj.

3 50k 1
2

1N-34A

pulse width adj.

3 100K 1
2

+12 V.D.C.

+12 to 25V.D.C. (The IRF640 will handle up to 200V.D.C. for when more exciting applications are desired.)

To lamps or other loads

1 Amp Fuse

5 Amp. Fuse

5 Amp Meter

S.P.S.T. SWITCH 6.0A. MIN.

Secondary

1000uF. ELECT.

47uF. TANTALUM

Primary

S.P.S.T. SWITCH

33uF TANTALUM CAP. 25V.

14 13 12 11 10 9 8

CD4584 or CD40106

HEX SCHMITT TRIGGER

1 2 3 4 5 6 7

.1 uF

8 7 6 5

TC4429CPA

HIGH SPEED CMOS DRIVER

1 2 3 4

POWER MOSFET Harris IRF640 etc.

Pot Pin-out

Rear View

3 2 1

DRAIN FLANGE

SOURCE
DRAIN
GATE

IRF640

POWER MOSFET Pin-out

By:Jim Campos

hex Schmitt trigger pulse generator

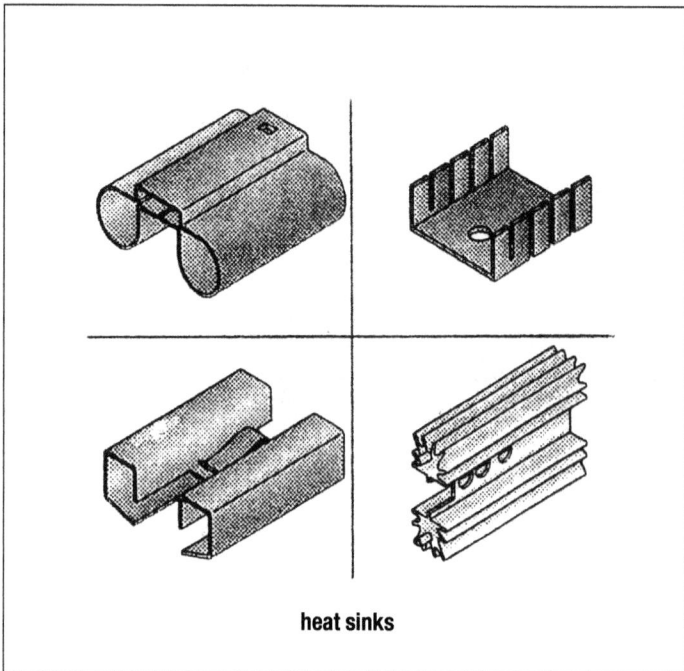

heat sinks

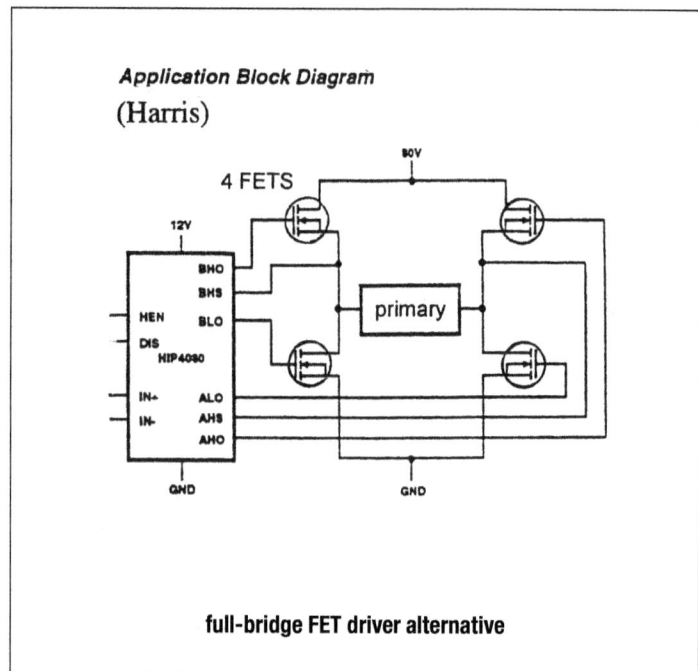

Application Block Diagram
(Harris)

full-bridge FET driver alternative

Series battery hook-ups of 24, 36, and 48 volts (possible in some home-power situations) produce superior results and efficiencies. The 12 to 2000-volt step-up itself represents a magnification of 166. Operating draw for the 12-volt lamp driver was 450MA to 1.5 amps, depending on the brightness desired. That's 5.4 to 18 watts.

fighting heat

To keep the "on" time at a minimum, the pulse must have a very sharp rise time and fall time, a sharply square wave. Any "ramping" will extend the "on" time, hence more heat. We fixed a heating problem that plagued us in the FET and FET driver as well when we changed FET drivers from a slower, amplifying 7667, which we used through many experiments, to the fast switching 4429.

When the circuit is in the "on" state conducting through the primary, its condition quickly becomes, after passing the 1/4-wave mark, a dead short. Momentarily, currents in excess of 18 amperes may be coursing through the power FET. At one megacycle and a quarter-wave pulse width, this short circuit would exist for nine percent of the time. The higher the frequency, the more of these short-circuit episodes in a given period of time. Heat builds and can only be dissipated at a limited rate.

We did a series of experiments to determine a safe operating frequency range in which this device could run cool and problem-free indefinitely. The pulse generator now operates comfortably driving Tesla coils at 300 kc and below.

Another good reason for a lower frequency of operation is that this level control by pulse-with control diminishes as the pulse width shrinks at higher frequencies.

If the circuit is pushed beyond the lighting-plant application here, you may want to install a second FET, using the second output of the FET driver.

alternatives

There are other ways to drive power transistors besides the 555/4584-4429 combination, and you may want to explore them. Many months of experimentation went into the development of the recipe circuit, however, and any new direction may require a similar investment. A possibility that has Jim's interest substitutes for the 555/4584 a quad-D flip-flop, which, in his scheme, would automatically provide for any given frequency a pulse width of quarter-wave duration.

An alternative to the 4429 of our recipe circuit may well be the 4080 full-bridge FET driver. The circuit shown is from a typical-application schematic in a reference source. The full ordering number for this chip is HIP4080AIP. It is more expensive than the 4429 arid harder to find. An Internet posting on the 4080 by a pyrotechnically oriented Tesla-coiler referred to "sheets of flame."

construction

Locate the big 1000 mfd. electrolytic close to the battery-supply terminal. Battery current will fill this big capacitor and it will hold plenty, ready for release. Right next to it put the 100 mfd. tantalum in parallel. This faster capacitor will release the energy in quick spurts. **HAZARD NOTE:** both of these capacitors are polarized, one lead is plus, the other minus. Connect them

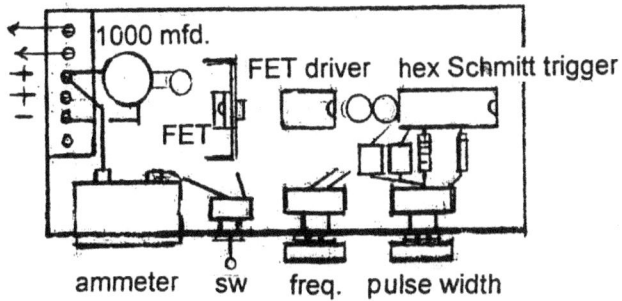

recipe pulse generator
suggested layout

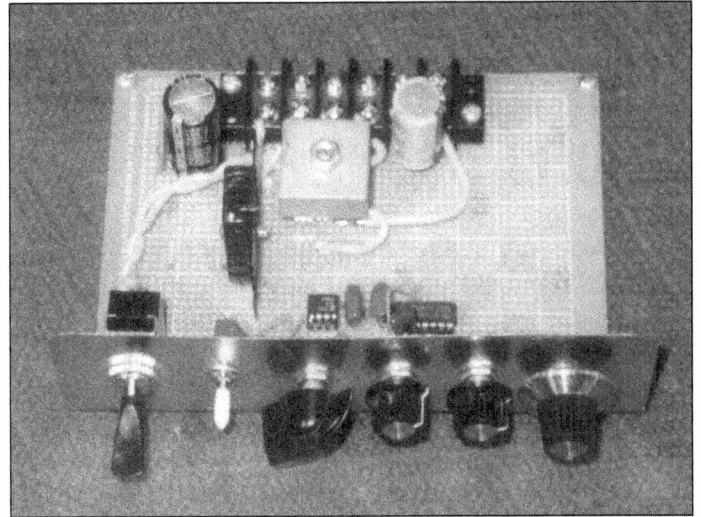

basic pulse generator

above, below: elaborated pulse generator

backwards and you get an explosion. Particularly dangerous is the tantalum, which becomes a little fragmentation bomb. Make sure the voltage rating of these caps matches your input voltage.

In the interest of expedient current delivery, use heavy wire, #18 or bigger (stranded, insulated) throughout the business end of the circuit: from the supply terminal to the heavy-duty 6-amp power switch, to the source and drain of the FET, and through to the primary. Use heavy wires to the unit from the battery, #14 or bigger. You may want to breadboard the circuit and get the feel of it before hardwiring. Stray capacitance can distort a pulse, so keep wires short. The little tantalum buffer caps should be snug up against their respective chips.

Parts Guide

Hard-to-find parts are specified here. See the schematic for such common items as pots, resistors, etc. Hex Schmitt trigger CD4584 (Harris), also CD40106 or NTE23740, from Mouser, or 555 timer from Mouser or Radio Shack. FET driver TC4-429CPA (Telecom) from Digikey. Power FET IRF 640, 200 volts, 18 amps (Harris), from Mouser, or lower power IRF 510 from Radio Shack. Tantalum caps: 68uf, 25-volt or 33uf, 25 volt from Mouser. Ammeter, 0-5 amps, Mouser. Mouser (800) 346-6873, Digikey (800) 344-4539.

CHAPTER 5

Build a Tesla Lighting Plant

If we use the solid-state pulse generator to drive a suitable Tesla coil, can we drive one or more fluorescent tubes at current-draw efficiencies superior to conventional fluorescents? I have done some experimenting suggesting that the answer to this question is yes. Experimenting also suggests that a design like this battery electronic Tesla coil could function coolly over long duty cycles, that it could function quietly and emission-free and require little maintenance. Properly installed, it should be safer in terms of shock hazard than conventional 120 VAC devices and safer probably in respect to fire hazard as well.

Although the prototype illustrated is far from compact, it's obvious that this criterion could be fulfilled, too. Yes, experimenting does suggest these potentialities for such a system, a Tesla Lighting Plant.

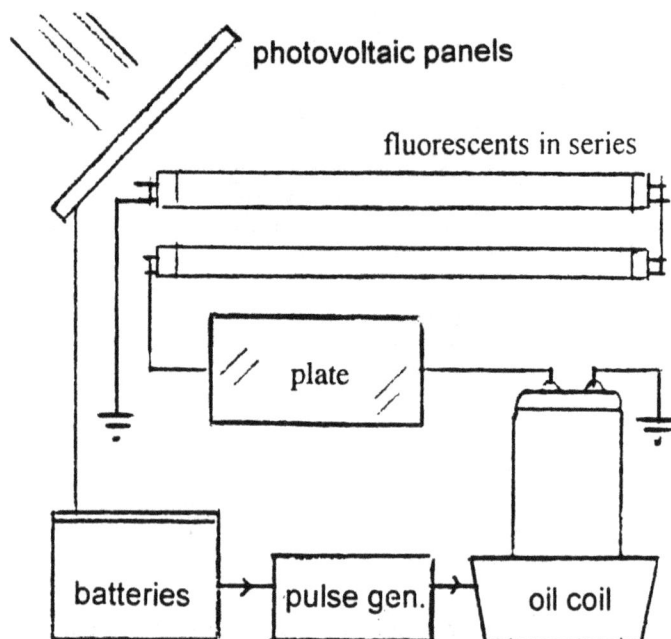

experiments

I soldered the two connecting wires directly to the fluorescent tube's end-pins, shorting each pair out. If you don't want to solder, use conventional fluorescent-tube sockets, which probably can be adapted for the purpose and will also serve to support the tube. Otherwise the tube can be lightly suspended in any well insulated fasliion. Another way "into" the tube that I experimented with was via bands of aluminum foil wrapped around the glass, a capacitive connection; but a direct connection to the pins works better.

My "plate" was just aluminum foil glued on to some 1/8-inch plastic sheet. The plate functions the same as a terminal capacitor (sphere or toroid) on a Tesla coil, raising up the voltage (and also incidentally lowering the frequency). The tube itself contributes some to the terminal capacity, but for best results this should be augmented by the plate. Like a flywheel in a mechanical system, some inertial load is needed for the Tesla coil to work on. In a practical appliance, the plate could be incorporated as a reflector, a shade, or a diffusor. The plate is a cheap way to kick up the voltage, but too much terminal capacity could drive the frequency down to where the system cannot be tuned.

For current-draw measurements I used a big laboratory ammeter set to the 0-3 amp scale. An ammeter is essential for tuning this system to efficiency arid would be incorporated into any ultimate commercial device as a panel meter (0-5 amp is commonly available). Exact voltage measurements, necessary to compute wattage, were made on a digital multimeter (DMM). The same DMM has a frequency-counter feature that was put to use, and I also used it for a crude light meter by plugging into it a cadmium sulfide photoresistor and setting the DMM for resistance measurement.

experimental limits

We devoted considerably more time to the development of the pulse generator and the Tesla coil than in developing the ideal way to connect the system to florescent tubes for peak performance and amp economy. I suspect there are discoveries ahead, and you may make some in your own experimenting. For example, in the interest of that fly-wheel effect, we have put a high-voltage capacitor across the output terminals of the Tesla coil and succeeded in driving the tube to a brilliance more than twice that of the conventional. Current draw was excessive in this experiment, but it was suggestive. What would variable capacity do? Also, experiment with various capacities placed across the primary coil in order to better tune that circuit.

Considerably more experimenting has been done with a single tube at 12 volts than with multiple tubes at higher inputs. Finally, experimenting has been limited to lighting, while I suspect that this electronic Tesla coil may have other practical applications, such as in ozone production or even heating or motive power.

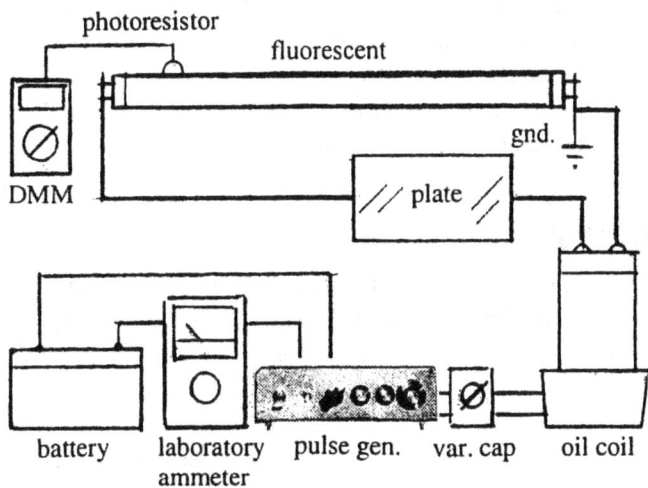

experimental set-up

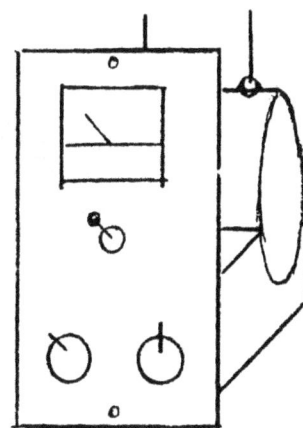

Tesla lighting plant
commercial concept

current economies

The pulse generator's internal chip electronics consumes a negligible 22 milliamperes. The ammeter measures the current of interest which is through the FET-Tesla-coil primary circuit. On the pulse generator you manipulate the frequency and pulse-width controls to produce the desired level of light at the minimum current draw. Check for excessive heat in the FETs and chips as you experiment. If current draw is held down to one ampere or less, the FETS will run cool and unstressed and long running times are possible. If the tube does not light up completely but glows at the ends, wait, and it will usually snap on bright. Touching the tube sometimes helps.

In tuning, you'll discover that an increase in light output does not necessarily require a commensurate increase in current consumption. In conventional lighting, an increase in light output always means more current, but here that rule is suspended. In fact, as you adjust frequency, there are settings where you'll see the amps dip as the lamp brightens. This is the dip familiar to the radioman tuning his transmitter, the dip on his milliammeter that tells him that the rig is operating on the money, the resonant dip.

A typical operating frequency in these experiments was 115kc.

At 12 volts and one tube it is possible to produce low-level practical room illumination, light suitable for walking around, drawing 450 MA or 5.4 watts. Turning it up toward adequate reading light level by advancing the pulse width, it's pulling 700 MA or 8.4 watts. At one amp the light level is about half that of a conventional tube while pulling one-third the power, 12 watts.

At 24 volts a single fluorescent tube starts up more easily and burns more brightly. Connect multiple tubes in series. It is in multiple-tube lighting that the advantages of this sort of system become more dramatic. In conventional fluorescents, a second tube means, inescapably, twice the wattage is used, a third tube, three times, etc. But in the Tesla system, this equation dissolves. One amp of input lights three tubes all to the same level as one amp lit one tube. The single tube lit at one amp, and 12 volts equaled 12 watts. The three tubes at 24 volts are together pulling one amp, which equals 24 watts, but now pulls only 8 watts per tube. Projected out a number of tubes this efficiency could amount to a considerable conservation of battery power.

Fluorescents lit Tesla-style can be adjusted for light level with the pulse-width control. Conventional fluorescents operate at a fixed output, are overly bright for some situations and cannot be mellowed out. Light levels can be attenuated for particular needs, resulting in current economies. Fluorescent tubes lit Tesla-style are healthier. The conventional fluorescent flickering at 60 cycles is irritating to the nervous system and can cause eyestrain and headaches. At 100 kc no flicker is perceptible.

space heating?

Wouldn't it be great to invent a Tesla-style battery powered space heater that pulls just a few amps? I don't have the answer, but I suspect that it would lie in the realm of vacuum technology. Tesla patented some high-potential, high-frequency lamps that contained solid elements in vacuum globes. Solid elements like little blocks of carbon burn brightly, and at high voltages reach such temperatures as to cause the solid element to fuse or vaporize. Could this phenomenon be translated into space heating?

growing

The only serious indoor growing operation I have observed close-up utilized an array of standard, cheap cool-white four-foot

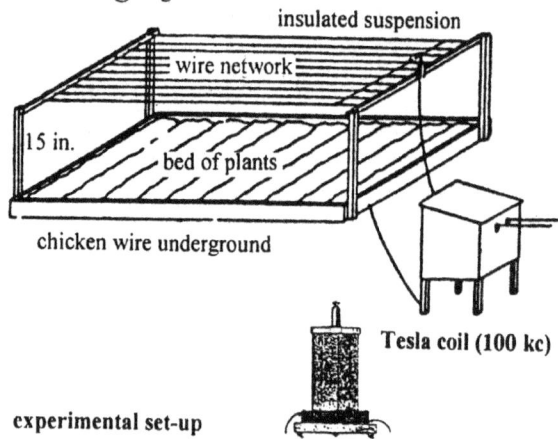

From *H.F Apparatus*
Curtis (1920). Thanks to Rex Research
rexresearch@biogate.com

insulated suspension

wire network

15 in.

bed of plants

chicken wire underground

Tesla coil (100 kc)

experimental set-up

high-frequency plant-growth stimulation

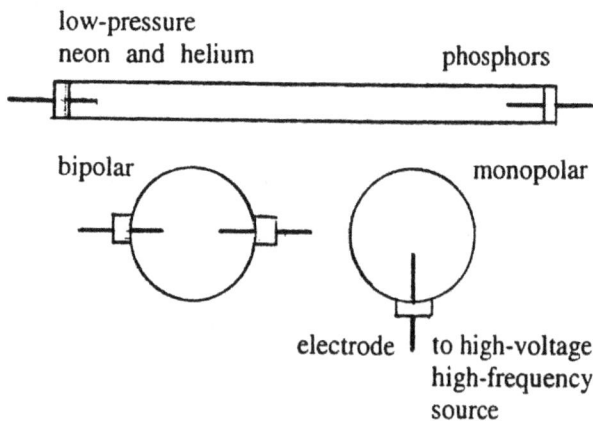

low-pressure
neon and helium   phosphors

bipolar   monopolar

electrode   to high-voltage
high-frequency
source

ideal Tesla lamp

There is scientific evidence that suggests that plants exposed to high-frequency, high-potential fields experience an accelerated rate of growth. High electric potentials at conventional 60-cycle frequencies evidently have a positive effect judging by the number of nurseries located under high-tension power lines, but high frequency may be superior.

safety

Particularly if driven by battery, the shock hazard of the system is negligible. Although the voltage at the Tesla-coil output end of the system is in the thousands, the frequency, being a hundred kc or more, does not register on the nervous system, which cuts off at about 3 kc. It is possible to touch bare terminals without sensation or harm. There is, however, a possibility of getting an rf burn if your finger tip serves to complete a circuit.

For fire safety (as well as to reduce electrical losses) it is a good idea to wire up the lamps with a very well-insulated conductor, such as 5 kv test wire. Poorly insulated wires that intersect may conceivably arc to one another causing a fire hazard. If you are making a long run from Tesla coil to lamps, keep wires that are in parallel a good distance apart. As more tubes are added and more input voltage supplied to the system, the output voltages can reach exceedingly high values making insulative anti-arcing and anti-coronal measures even more important.

radio interference

The conventional florescent tube, flickering at 60 cycles, produces wide-band disturbances which are carried over power lines and can blot out all but strong local stations over large segments of the AM broadcast band. The Tesla system can produce at discreet points on the band little tones and squeals from harmonics. These fade off quickly with distance from the source and are not carried over the power grid. Output at the fundamental frequency is assumed to be translated into light, the tube serving, as it were, as a dummy load. Any other unwanted oscillations could conceivably be subdued by filtration.

the ideal appliance

The ubiquitous cool-white notwithstanding, what would be the ideal lamp for Tesla currents were one to invent it from scratch? I dropped this very question upon Jim Hardesty, whose PV Scientific produces elegant Crookes tubes used to demonstrate vacuum phenomena. His suggestion: Fill a tube with a mixture of helium and neon at low pressure, say .1 mm of mercury. One can experiment with a vacuum pump on an electrified tube to determine the optimal pressure.

A clear glass tube can be used, but, by experimenting with different phosphors as interior coatings, one can produce double or triple the amount of light for the energy put in.

fluorescent tubes. The grower, who could have afforded the pricier grow lights or even arc lamps, claimed that the old cool-white was his lamp of choice. I will vouch for the quality of his crop. His lamps, of course, are the same cool-whites that abound, in a technically burnt-out condition, in the dumpsters of office and apartment buildings everywhere waiting to get a new life by feeding on high-potential, high-frequency Tesla currents.

Fluorescents driven Tesla-style run cool. No cathodes are burning. The Tesla-current itself produces no heat that could be detectable. Fluorescents driven Tesla-style may deliver current economies that would allow operation from solar-charged battery sources rather than noisy and potentially flammable gasoline generators.

CHAPTER 6

Designing Radio Tuning Tanks

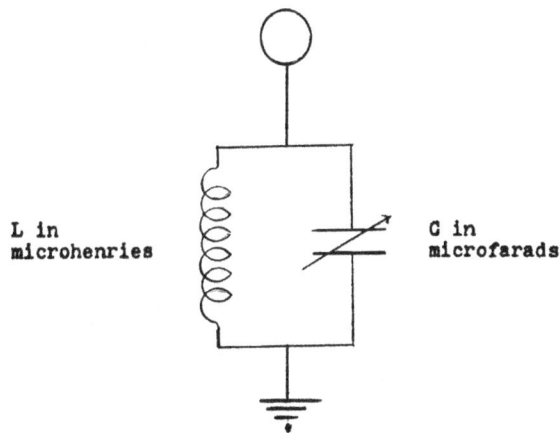

The simple tank circuit is one of the core secrets in the magic of radio. It's true that in the higher frequencies explored later in the history of radio, the tuning tank becomes problematic; this happens in the range of 10 to 40 megacycles. But for any frequency below that (in the shortwave, broadcast AM, or longwave bands) all you need for the frequency-determining circuitry of any receiver or transmitter is one or more resonant tuning tanks.

To calculate the appropriate values for the coil and capacitor of a tank circuit so that, together, they resonate at a desired frequency, first calculate the wavelength: Divide the frequency (expressed in cycles per second) into 300,000, the approximate speed of light in meters per second. This gets you the wavelength in meters. Then you can use this old formula:

$$wavelength = 1885\sqrt{LC}$$

Or, solving for LC, you get the practical formula:

$$LC = \left(\frac{wavelength}{1885}\right)^2$$

L is the inductance in microhenries, C is the capacity in microfarads, and the LC is the inductance times the capacity. If you know the value of one, you can find the other by dividing the known into the LC number given by the above equation, a value sometimes called the oscillation constant.

If you would rather not do the oscillation constant calculations, use the Tank Tables below. Usually the variable capacitor will be the known and the coil the unknown. Variables are usually rated in picofarads; move the decimal point six places to the left to get microfarads. Be sure your inductance is in microhenries not millihenries.

The texts say that for high Q there should be more capacitance than inductance in a tank. In designing a coil, diameter and length of winding should be about equal.

A formula that translates a single-layer air-core coil's physical specs into microhenries is

$$L = \frac{(N \times A)^2}{9A + 10B}$$

where L is the coil's inductance, N is the number of turns of wire, A is the length of the coil winding, and B is the diameter, both in inches. Divide the oscillation constant for the frequency you want by the capacitance in microfarads and you'll get the

inductance in microhenries that you can design the coil for. By transposing the coil formula, you can yield the number of turns. By plugging in lengths and diameters, the physical characteristics of a coil with the desired inductance will begin to emerge.

$$N = \sqrt{\frac{L(9A + 10B)}{A}}$$

quarter-wave

Tesla called his quarter-wave principle "the secret of tuning." He said that "without the observation of this rule it is impossible to prevent the interference and insure the privacy of messages." Tesla's quarter-wave formula is still honored in calculating the length of antenna elements and is still useful in calculating an antenna-loading coil and for tuning circuits that have a coil alone without a capacitor.

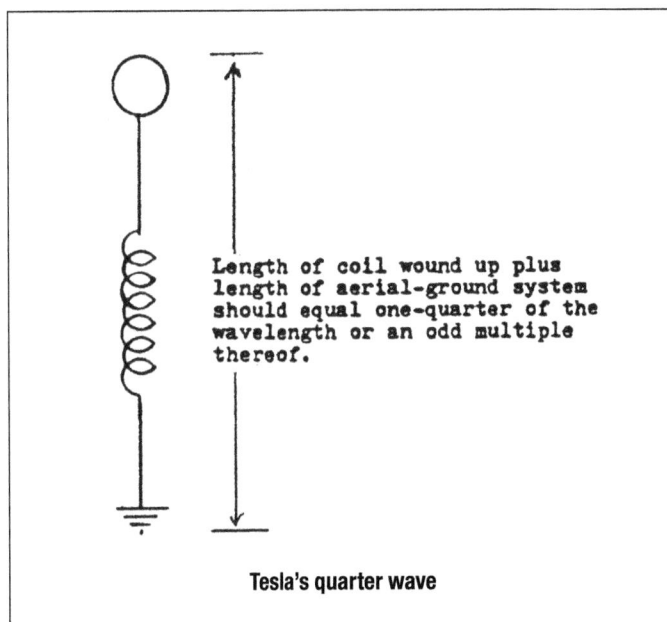

Length of coil wound up plus length of aerial-ground system should equal one-quarter of the wavelength or an odd multiple thereof.

Tesla's quarter wave

Tank Tables
wavelength, frequency and oscillation constant

WAVELENGTH, FREQUENCY, AND OSCILLATION CONSTANT

Wave-length Meters	Frequency	LC	Wave-length Meters	Frequency	LC
1	300,000,000	.0000003	190	1,579,000	.01016
2	150,000,000	.0000011	195	1,538,000	.01071
3	100,000,000	.0000025	200	1,500,000	.01126
4	75,000,000	.0000045	205	1,463,000	.01183
5	60,000,000	.0000070	210	1,429,000	.01241
6	50,000,000	.0000101	215	1,395,000	.01301
7	42,860,000	.0000138	220	1,364,000	.01362
8	37,500,000	.0000180	225	1,333,000	.01425
9	33,333,000	.0000228	230	1,304,000	.01489
10	30,000,000	.0000282	235	1,277,000	.01555
15	20,000,000	.0000634	240	1,250,000	.01622
20	15,000,000	.0001126	245	1,225,000	.01690
25	12,000,000	.0001760	250	1,200,000	.01760
30	10,000,000	.0002533	255	1,177,000	.01831
35	8,571,000	.0003448	260	1,154,000	.01903
40	7,500,000	.0004503	265	1,132,000	.01977
45	6,667,000	.0005700	270	1,111,000	.02052
50	6,000,000	.0007039	275	1,091,000	.02129
55	5,454,000	.0008519	280	1,071,000	.02207
60	5,000,000	.001014	290	1,034,500	.02366
65	4,615,000	.001188	295	1,017,000	.02450
70	4,286,000	.001378	300	1,000,000	.02533
75	4,000,000	.001583	310	967,700	.02705
80	3,750,000	.001801	320	937,500	.02883
85	3,529,000	.002034	330	909,100	.03066
90	3,333,000	.002280	340	882,400	.03255
95	3,158,000	.002541	350	857,100	.03448
100	3,000,000	.002816	360	833,300	.03648
105	2,857,000	.003105	370	810,800	.03854
110	2,727,000	.003404	380	789,500	.04065
115	2,609,000	.003721	390	769,200	.04277
120	2,500,000	.004052	400	750,000	.04503
125	2,400,000	.004397	410	731,700	.04733
130	2,308,000	.004757	420	714,300	.04966
135	2,222,000	.005130	430	697,700	.05204
140	2,144,000	.005518	440	681,800	.05446
145	2,069,000	.005919	450	666,700	.05700
150	2,000,000	.006335	460	652,200	.05960
155	1,935,000	.006760	470	638,300	.06219
160	1,875,000	.007204	480	625,000	.06485
165	1,818,000	.007662	490	612,200	.06759
170	1,765,000	.008134	500	600,000	.07039
175	1,714,000	.008620	510	588,200	.07327
180	1,667,000	.009120	520	576,900	.07606
185	1,622,000	.009634	530	566,000	.07905

1

WAVELENGTH, FREQUENCY, AND OSCILLATION CONSTANT—(Continued)

Wave-length Meters	Frequency	LC	Wave-length Meters	Frequency	LC
540	555,600	.08208	990	303,100	.2759
550	545,400	.08519	1,000	300,000	.2816
560	535,700	.08836	1,010	297,000	.2870
570	526,300	.09139	1,020	294,100	.2927
580	517,200	.09467	1,030	291,300	.2986
590	508,500	.09801	1,040	288,400	.3045
600	500,000	.1014	1,050	285,700	.3105
610	491,800	.1047	1,060	283,600	.3161
620	483,900	.1082	1,070	280,400	.3222
630	476,200	.1117	1,080	277,800	.3283
640	468,700	.1154	1,090	275,200	.3345
650	461,500	.1188	1,100	272,700	.3404
660	454,500	.1225	1,110	270,300	.3467
670	447,800	.1263	1,120	267,900	.3531
680	441,200	.1302	1,130	265,500	.3595
690	434,800	.1341	1,140	263,100	.3660
700	428,600	.1378	1,150	260,900	.3721
710	422,500	.1419	1,160	258,600	.3787
720	416,700	.1459	1,170	256,400	.3853
730	411,000	.1501	1,180	254,200	.3921
740	405,400	.1540	1,190	252,100	.3988
750	400,000	.1583	1,200	250,000	.4052
760	394,800	.1626	1,210	247,900	.4121
770	389,600	.1668	1,220	245,900	.4190
780	384,600	.1712	1,230	243,900	.4260
790	379,800	.1756	1,240	241,900	.4326
800	375,000	.1801	1,250	240,000	.4397
810	370,400	.1847	1,260	238,100	.4469
820	365,900	.1893	1,270	236,200	.4541
830	361,400	.1941	1,280	234,400	.4610
840	357,100	.1985	1,290	232,600	.4683
850	352,900	.2034	1,300	230,800	.4757
860	348,800	.2082	1,310	229,000	.4831
870	344,800	.2132	1,320	227,300	.4906
880	340,900	.2179	1,330	225,600	.4978
890	337,100	.2229	1,340	223,900	.5053
900	333,300	.2280	1,350	222,200	.5130
910	329,700	.2332	1,360	220,600	.5208
920	326,100	.2381	1,370	218,900	.5281
930	322,600	.2434	1,380	217,400	.5359
940	319,100	.2487	1,390	215,800	.5438
950	315,900	.2541	1,400	214,300	.5518
960	312,500	.2595	1,410	212,800	.5598
970	309,300	.2647	1,420	211,300	.5674
980	306,100	.2704	1,430	209,800	.5755

2

WAVELENGTH, FREQUENCY, AND OSCILLATION CONSTANT—(Continued)

Wave-length Meters	Frequency	LC	Wave-length Meters	Frequency	LC
1,440	208,300	.5837	1,890	158,700	1.006
1,450	206,900	.5919	1,900	157,900	1.016
1,460	205,500	.5998	1,910	157,100	1.027
1,470	204,100	.6081	1,920	156,300	1.038
1,480	202,700	.6165	1,930	155,400	1.049
1,490	201,300	.6250	1,940	154,600	1.060
1,500	200,000	.6335	1,950	153,800	1.071
1,510	198,700	.6416	1,960	153,100	1.081
1,520	197,400	.6502	1,970	152,300	1.092
1,530	196,100	.6590	1,980	151,500	1.104
1,540	194,800	.6677	1,990	150,800	1.115
1,550	193,600	.6760	2,000	150,000	1.126
1,560	192,300	.6849	2,050	146,300	1.183
1,570	191,100	.6938	2,100	142,900	1.241
1,580	189,900	.7028	2,150	139,500	1.301
1,590	188,700	.7118	2,200	136,400	1.362
1,600	187,500	.7204	2,250	133,300	1.425
1,610	186,300	.7295	2,300	130,400	1.459
1,620	185,200	.7387	2,350	127,700	1.555
1,630	184,100	.7480	2,400	125,000	1.622
1,640	182,900	.7573	2,450	122,500	1.690
1,650	181,800	.7662	2,500	119,000	1.760
1,660	180,700	.7756	2,550	117,700	1.831
1,670	179,600	.7852	2,600	115,400	1.903
1,680	178,600	.7946	2,650	113,200	1.977
1,690	177,500	.8037	2,700	111,100	2.052
1,700	176,500	.8134	2,750	109,100	2.129
1,710	175,400	.8231	2,800	107,100	2.207
1,720	174,400	.8329	2,850	105,300	2.287
1,730	173,400	.8422	2,900	103,500	2.366
1,740	172,400	.8520	2,950	101,700	2.450
1,750	171,400	.8620	3,000	100,000	2.533
1,760	170,500	.8720	3,100	96,770	2.705
1,770	169,400	.8821	3,200	93,750	2.883
1,780	168,500	.8916	3,300	90,910	3.066
1,790	167,600	.9018	3,400	88,240	3.255
1,800	166,700	.9120	3,500	85,910	3.448
1,810	165,700	.9223	3,600	83,330	3.648
1,820	164,800	.9327	3,700	81,080	3.854
1,830	163,900	.9425	3,800	78,950	4.065
1,840	163,000	.9530	3,900	76,920	4.277
1,850	162,200	.9634	4,000	75,000	4.503
1,860	161,300	.9741	4,100	73,170	4.733
1,870	160,400	.9841	4,200	71,430	4.966
1,880	159,600	.9948	4,300	69,770	5.204

3

WAVELENGTH, FREQUENCY, AND OSCILLATION CONSTANT—(Continued)

Wave-length Meters	Frequency	LC	Wave-length Meters	Frequency	LC
4,400	68,180	5.446	8,800	34,090	21.79
4,500	66,670	5.700	8,900	33,710	22.29
4,600	65,220	5.960	9,000	33,330	22.80
4,700	63,830	6.219	9,100	32,970	23.32
4,800	62,500	6.485	9,200	32,610	23.81
4,900	61,220	6.759	9,300	32,260	24.34
5,000	60,000	7.039	9,400	31,910	24.87
5,100	58,820	7.327	9,500	31,590	25.41
5,200	57,690	7.606	9,600	31,250	25.95
5,300	56,600	7.905	9,700	30,930	26.47
5,400	55,560	8.208	9,800	30,610	27.04
5,500	54,550	8.519	9,900	30,310	27.59
5,600	53,570	8.836	10,000	30,000	28.16
5,700	52,630	9.139	10,500	28,570	31.05
5,800	51,720	9.467	11,000	27,270	34.04
5,900	50,850	9.801	11,500	26,090	37.21
6,000	50,000	10.14	12,000	25,000	40.52
6,100	49,180	10.47	12,500	24,000	43.97
6,200	48,550	10.82	13,000	23,080	47.57
6,300	47,620	11.17	13,500	22,220	51.30
6,400	46,870	11.54	14,000	21,440	55.18
6,500	46,150	11.88	14,500	20,690	59.19
6,600	45,450	12.25	15,000	20,000	63.35
6,700	44,780	12.63	15,500	19,350	67.60
6,800	44,120	13.02	16,000	18,750	72.04
6,900	43,480	13.41	16,500	18,180	76.62
7,000	42,860	13.78	17,000	17,650	81.34
7,100	42,250	14.19	17,500	17,140	86.20
7,200	41,670	14.59	18,000	16,670	91.20
7,300	41,100	15.01	18,500	16,220	96.34
7,400	40,540	15.40	19,000	15,790	101.64
7,500	40,000	15.83	19,500	15,380	107.06
7,600	39,470	16.26	20,000	15,000	112.56
7,700	38,960	16.68	21,000	14,290	124.12
7,800	38,460	17.14	22,000	13,640	136.24
7,900	37,980	17.56	23,000	13,040	148.93
8,000	37,500	18.01	24,000	12,500	162.15
8,100	37,040	18.47	25,000	12,000	175.97
8,200	36,590	18.93	26,000	11,540	190.26
8,300	36,140	19.41	27,000	11,110	205.20
8,400	35,710	19.85	28,000	10,710	220.70
8,500	35,290	20.34	29,000	10,350	236.63
8,600	34,880	20.82	30,000	10,000	253.32
8,700	34,480	21.32			

4

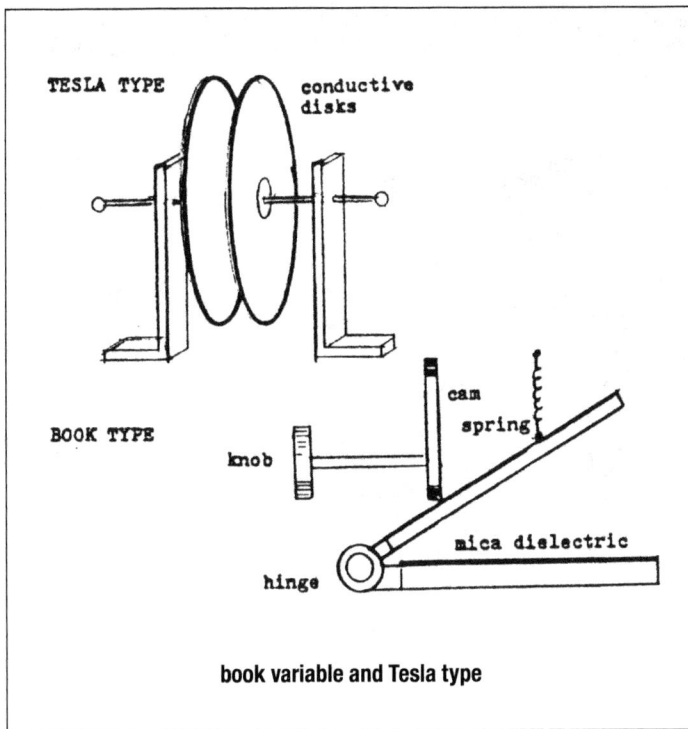

book variable and Tesla type

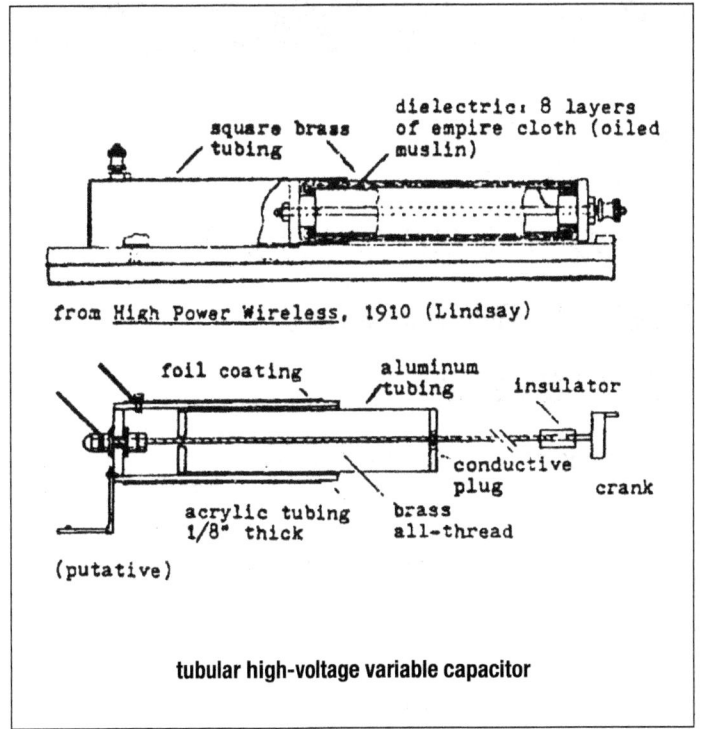

from High Power Wireless, 1910 (Lindsay)

(putative)

tubular high-voltage variable capacitor

building variable capacitors

You can build an adjustable salt-water capacitor out of beer bottles, as shown above in *Build a Tesla Coil*. Old circuits show variable capacitors consisting simply of a cluster of fixed capacitors and a rotary switch. Tesla's variable capacitor consisted of two opposing conductive disks whose distance apart was adjustable. Transmitters sometimes use little neutralizing variables of this design, but I've seen it nowhere else. The book type was manufactured by Crosley in the 1920s. The tubular shows up in the transmitter literature circa 1910. Even the popular rotary-plate type of variable capacitor is becoming extinct, being almost completely out of manufacture as new tuning methods (varactor diodes, frequency synthesizers) have come into use.

Mini-variables, almost useless for transmitting, are more easily found.

Especially hard to find are the transmitter variables which have widely spaced plates to take high voltage. The rotary-plate variable is a sophisticated piece of machinery that is difficult to build yourself, so it is becoming useful to know about the old book, tubular, and Tesla types, which are relatively easy to construct.

Variable capacitors include the familiar rotary plate (also called "air variable" because the dielectric is air), the simply constructed "book" type, and the tubular high-voltage type for transmitting.

CHAPTER 7

Build a Low-Frequency Regenerative Receiver

author's low-frequency regen

The project shown is an adaptation of a 1930's vacuum-tube regen to a circuit using a single field-effect transistor. *The Official 1934 Shortwave Manual* offers a similar circuit in an appendix written by the publisher (Lindsay). I've modified it for longwave. Here is an inexpensive way to go on a fishing expedition in a segment of the low-frequency band.

The project is built on a wooden chassis. The metal panel discourages hand-capacitance effects. My coil was wound for about 200 kc. Coils can be wound for other segments of the low band, and you can also wind coils for shortwave. High-impedance phones are necessary for this single-stager, but conventional earphones or even a loud-speaker can be driven if you add an audio amplifier. The 365 pf main tuning variable can be scrounged from an old broadcast receiver or ordered from Antique Radio Supply, which is also a source for shortwave plug-in coil forms as well as the high-impedance phones. The

author's regen rear view

neat little vernier dial (optional) enables slow, precise tuning of the main variable. The little bandspread variable (also optional) is a surplus or hamfest item.

In the conventional regenerative project, the tickler coil is wound on the same form as the tank and antenna coils. It requires lengthy trial-and-error to find the exact number of turns that will bring the receiver to life. After endless hours fiddling with a conventional tickler, I improvised a coil that would fit inside the main coil and was movable by small increments. Easily, I found the single sensitive position that triggered the circuit into regeneration. Later I discovered in the literature that my "invention" had once been the customary mode. The regen had gone into mass production with fixed ticklers; and the movable had gone out of fashion even among those who built their own, and was forgotten.

I glued strips of rubber on the tickler's form so it fitted snugly inside. (An ideal arrangement would be to mount the tickler so it rotates within as a variometer. Same for the antenna coil; it would add another dimension of tuning.) The pot controls both the volume and sensitivity (squelch) by controlling the degree of regeneration. Too much regeneration can send the set into howling oscillation, but first try the set with the control full on. Tune for a station to give the set some energy to work with. And, yes, you have good ground and at least 30 feet of long-wire antenna or equivalent. An antenna-tuning coil helps a lot. The regen wants a loose coupling to the antenna. You can insert another variable capacitor in the antenna circuit.

If you think you've done everything right and have double-checked the wiring, and the set still has not come to life, reverse the leads to the tickler coil. With a vacuum-tube regen sometimes an otherwise OK tube refuses to oscillate, and a replacement is called for. This may apply also to transistors. Positioning of the tickler is critical. Touch the antenna side of the tank coil, if you hear a click, this is a sign of life.

This advice smugly given, I'll add that the regen operates in mysterious ways, and, especially in the process of bringing it to life, may try your patience. It may also give you some surprises. I once stuck a potent little bar magnet inside the coil at the critical regeneration point and it burst into incredibly loud oscillation.

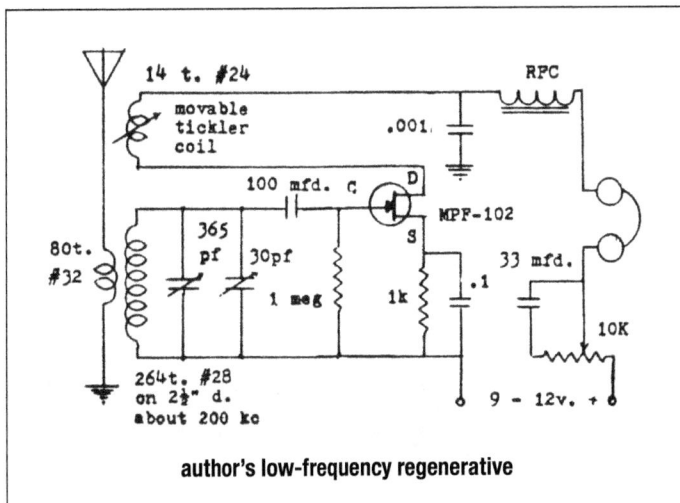

author's low-frequency regenerative

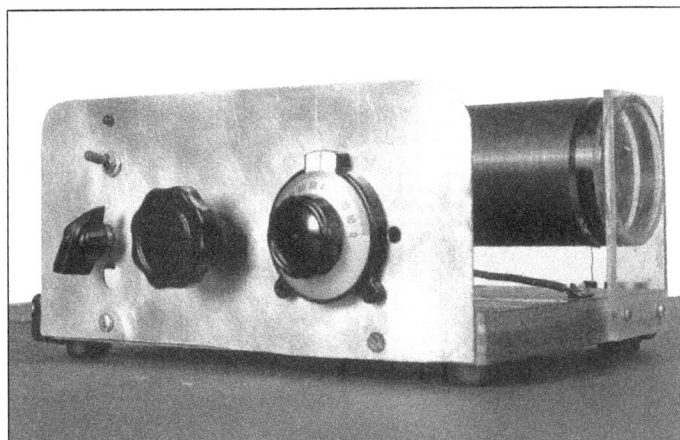

With the magnet removed, the coil continued to function strangely, triggering too easily into howling, and had to be set aside.

My low-frequency regen had considerable difficulty pulling in more than one beacon while operating it in the city (Portland, OR). There was too much electrical noise and scandalously abundant longwave (and shortwave) harmonics from sloppy AM broadcast transmitters. When I took this set to Waldron Island, WA, where there is no electric power or nearby transmitters, it immediately pulled in 23 marine beacons.

Appendix C

Colorado Springs

by Nikola Tesla

edited by George Trinkaus

from Tesla's Colorado Springs notes 1899 – 1900

August 2, 1899 is a most unusual entry in Tesla's Colorado diary, which consists mainly of brief notations on magnifying-transmitter phenomena. Tesla spent just a year in Colorado, beginning in May, and he wrote this entry in August, so he writes almost as a tourist. The essay is a showcase for Tesla the writer/observer. I call it an aerial travelogue, because Tesla shows little interest in the earthly landscape ("barren") and devotes only a few paragraphs to the local people ("mostly consumptives". Consumption is the 19th-century term for tuberculosis.) Tesla speculates scientifically on why these patients benefit so much from the Colorado Springs environment. Bored with the mundane, Tesla, the newcomer, looks upward with astonishment at the drama of the Colorado Springs skies.

— GT

An Aerial Travelogue

August 2, 1899

First of all one is struck by the extraordinary purity of the atmosphere, which is best evident from the clearness and sharpness of outlines of objects at great distances. In low regions, especially where moisture is in excess, the outlines of objects become more or less indistinct and confuse at distances of but a very few miles, while here the outlines appear perfectly clear and sharp. When a train is moving up Pike's Peak, it is quite easy to distinguish, not only the engine and cars, but even the windows and wheels perfectly, although the distance from the experimental station is about ten miles. Quite frequently the house on top of Pike's Peak can be clearly seen with the naked eye. The ranges of mountains 100 to 150 miles away or more can be perceived perfectly. A range at a distance of about 50 miles can be seen plainly even at night when the sky is clear. It is wonderful how at times immense objects appear dwarfed, while small objects as horses, carriages or men assume unnatural gigantic dimensions.

Pike's Peak Range appears at times so close and so ridiculously small, that anyone not knowing the reality would be apt to fire a modern rifle at some object on the mountain-side believing it to be within shot. At other times Pike's Peak appears far remote and its height much beyond what would seem natural. The arc lamps at the foot of the mountains five to seven miles away shine with a brilliancy as though they were only as many blocks from the observer, and under certain conditions an ordinary incandescent lamp of 16 candlepower seems to give out as much light as an ordinarily arc light does. It appears also as big as the latter. This penetration of the light is due to the wonderful purity and extreme dryness of the atmosphere.

moon shadows

The moonlight is of a power baffling description. I have been told that the best photographs of the mountains have been obtained by moonlight, and I do not doubt it. Exposures of half an hour ought to give clear photographs revealing all details although the exposures are, as I am told, from one-and-a-half to two hours. I have nowhere seen such a light. Italy is famous for moonlight nights, but in my estimation that country cannot even compare with Colorado.

I think this extraordinary brightness of the moonlight is chiefly due to the absence of moisture, for there are many places, as in Central America, located much higher, and yet the moonlight, I am told, is not so intense, and I can see no other reason for this except the presence of more vapor in those places. It is not a mere saying, but literally true, that, during full moon in these parts, it is "as light as day." Objects can be clearly perceived at distances of many miles, and one can easily recognize a friend or familiar object at a distance of a quarter mile.

Colorado Springs, 1900
altitude 6000 feet, Pikes Peak 14,000

The shadows cast by the moonlight are extraordinarily black and sharp. They suggest the Crookes' shadows noted in vacuum bulbs, and, on this account, the moonlight is particularly interesting, suggesting thought and stimulating the imaginative powers. The shadows of the clouds on the plains and mountains are quite dark and clearly defined, and it is interesting to behold the patches as they speed over the ground.

brilliant stars

When the moon is absent and the nights clear, the number of stars visible and their brilliancy is amazing, and the sky presents a truly wonderful sight. The twinkling of the stars is very pronounced; they seem to move in orbits of as much as ten or fifteen of their own diameters across. At times one observes a star burst out into great brilliancy. This is probably due to the removal of an invisible cloud or of a layer of air at a great altitude containing some kind of particles which cut off a large portion of the light. One sees shooting stars quite frequently, also colored rings around the moon, generally in the advanced hours of the night, at times when the air is slightly misty. As this happens generally during very cold nights, I believe the colored rings are due to minute crystals of ice.

Owing to the extraordinary purity and dryness of the atmosphere the sounds penetrate to astonishing distances. This is particularly true of high notes. Certain conditions, entirely exceptional, concur at times and produce effects of this kind, which are startling. A bell will ring in the city several miles away, and it would seem to be before the very door of the laboratory. During certain nights when sleepless I have been astonished to hear the talk of people in the streets and sounds of this kind in a large radius around the dwelling, not to speak of the grinding of the wheels, the rolling of wagons, the puffing of the engines etc., which are perceptible in such a case, and with painful loudness though coming from distances incredibly great.

These phenomena are so striking that they can not be satisfactorily explained by any plausible hypothesis, and I am led to believe that possibly the strong electrification of the air, which is often noted, and to an extraordinary degree, may be more or less responsible for their occurrence.

kiln-dry

The dryness of the atmosphere, which is still further enhanced by the low pressure, is such that wood or other material is made what is called kiln-dry inside of a few hours and is rendered an insulator far more perfect than wood is ordinarily.

The nails on the hands and toes dry out to such an extent that they break off very easily, in fact one has to be careful in trimming them. I found the claws of a cat as brittle as glass. The skin on the hands dries out and cracks up and is apt to form deep sores particularly if, as often in experimentation, one has to wash the hands frequently. The hair gets perceptibly thinner owing to the drying out. Colorado is not a good country for hair. This may be of interest to people with a tendency towards baldness. People even very sick do not cough and expectorate, evidently owing to the dryness of the atmosphere. One does not perspire, as the sweat is immediately evaporated. It is curious how quick the body gets dry when a bath is taken. Still more this is noted when the body is rubbed with alcohol. These observations are not often made, unfortunately, as the opportunities for comfort are not such as one might desire.

the sky

In many respects one is disappointed with the aspect of the country itself, although it is far famed. I think it very uninteresting. Even the celebrated Pike's Peak is insignificant. Most of the country is barren, practically a desert, with little vegetable and animal life in places. Prairie dogs are about the only animals one can see on the plains. One rarely sees a bird, and the country must be a tedious one to live in for any one with tastes for hunting and fishing.

But as much as the country is devoid of interest and beauty, so much and far more is the sky beautiful. The sights one sees here in the heavens are such that no pen can ever describe. The cloud formations are the most marvelous sights that one can see anywhere. The iridescent colors are to my judgment incomparably more vivid and intense than in the Alps. Every possible shade of color may be seen, the red and white preponderating.

The phenomena accompanying the sunrise and sunset are often such that one is at the point of not believing his own eyes. At times large portions of the sky assume a deep red almost blood-red color, so intense that superstitious people might well be frightened when first seeing it, as by some other altogether unusual manifestation in the heavens.

Sometimes, particularly in the forenoon, huge masses of what appears to be snow are seen floating in the air, and they are so

Tesla's Colorado Springs lab

real and tangible, so sharply defined, that it is difficult to believe them to be composed merely of vapor. The purity and dryness of the atmosphere explains to a degree the sharpness of definition of the boundaries of these formations of mist, but it is quite possible that some other causes, as electrification of the particles, cooperate in rendering them so compact as they appear to be. Of course, the purer the air, the greater is the difference between the region filled by cloud and that surrounding it as regards the passage of light rays, and the boundaries of the cloud appear sharper and quasi-solid much on this account. The whiteness and purity of these masses of cloud is such that one has the idea that nothing, not even an angel, could come in contact with it without soiling it.

Very often, when the sun is setting, a considerable portion of the sky above the mountain range presents the sight of an immense furnace with white-hot molten metal. It is absolutely impossible to look at the melting-away clouds without being blinded, so vivid is the light. On a few occasions I have seen the mountains covered with a white silvery veil most beautiful to see, an unusual occurrence and caused by a fine mist-like rain in the mountain region. The intensity of the light on these occasions was really wonderful.

What was remarked before of the shadows of the moon is, and to a much greater degree, true of those thrown by the sun. They are ink-black and sharply outlined.

The shadows on the plain and mountains thrown by the clouds appear like big patches of inking blackness hurrying along the ground. Particularly interesting are shadows thrown across the sky resembling often large dark streamers, or those which under certain conditions are formed and are visible like dark columns extended from the ground to the sky. These shadows seem to be best visible in the middle of the afternoon or a little later when the sun in fairly down and on days when it has been extremely hot and sultry in the forenoon and the clouds are formed quickly and are of greater density than usual.

cloud-watching

A very curious phenomenon is the rapid formation and disappearance of the clouds. One can watch them continuously forming and disappearing rapidly, and one merely needs to turn away for a few moments when he may see that the aspect has changed, new clouds having replaced those he saw before. On many occasions, just after sunset, I have seen seemingly dense, white clouds appear as by enchantment below the mountain peaks. So quickly did these clouds or mist form that their appearance was much like the projection of an image on the screen.

The wonderful beauty of the cloud formations as seen here is, however, enhanced, not only by the incredible sharpness of the outlines and vividness of color, but also by their accidental arrangement and forms they assume. Not infrequently one can see clouds resembling all kinds of known objects, this adding much to the enjoyment one finds in observing them. In fact I have scarcely ever watched the clouds here without noting among them shapes resembling one or another familiar object. It is probably owing to the peculiar character of the clouds here that phenomena of this kind may be almost daily observed, whereas in other parts they are very rare.

Very often I have seen low on the horizon what appeared to be immense fields of ice as a sea frozen in the midst of a storm but so wonderfully real that it would be impossible to give an idea of it by a description however vivid. At other times there appeared ranges of mountains which one could not distinguish from the actual, on the horizon or the wide ocean, with its deep green, or dark blue, or black waters stretching out as far as the eye could reach. Nor was this an ordinary resemblance which one could banish from the mind by a small effort of will, but was rather of nature of those visions or hallucinations which make it necessary for one to pinch himself to fully realize that his senses have been deceiving him. More than once I have seen this ocean dotted with green islands or populated with glittering icebergs or sailing vessels or even steamers not less real to the eye because they were formations of mere mist or cloud.

Almost every evening, after sunset, and when the sky is clear, the horizon towards the plains becomes peculiarly tinged with colors of surprising vividness, all the colors of the rainbow being represented, the strata higher above the horizon beginning with red and passing through all nuances, the lowest strata finishing with blue, violet and black. As it grows darker the black line rises continuously above the horizon. This phenomenon illustrates in an interesting manner how the sun's rays are deviating from the straight course and are being continuously deflected downwards to the more dense strata of the atmosphere. Among the seemingly infinite variety of clouds there are four typical forms regularly observable which are of surpassing beauty. They are:

1. red clouds

Seen very frequently in the early morning hours at sunrise and, though less frequently, in the evening when owing to a greater percentage of moisture the clouds are denser, more like rain clouds. They reach an intensity of color equal to that of a ruby of the "pigeonblood" species. They are particularly beautiful when appearing in detached masses.

2. white clouds

Seen chiefly in the forenoon or in the early part of the afternoon though not so often. The whiteness and purity of these clouds and their sharpness of contours makes them a unique sight. It would be difficult to offer to the eye a greater treat than it finds in the contemplation of these masses of mist, generally floating in big detached lumps in the blue sky. I note that these clouds are seen generally after a short rain when the wind, springing up suddenly, clears the sky, leaving only a few large and separate masses of vapor.

3. clouds of gold

These present the appearance of immense lumps of gold. They are iridescent clouds witnessed chiefly at sunset. They present a striking sight, particularly when they are small and detached from each other and the sun's rays can penetrate them more freely, thus heightening at times, to a degree really incredible, the intensity of the iridescence. Their color is absolutely like that of gold and the similarity is rendered complete by the forms they assume which are those of gold nuggets found in nature, but generally they pass from pure yellow to a reddish yellow of the kind peculiar to gold found in certain countries or generally gold containing a small percentage of copper. A feature of these most beautiful clouds is that they persist in their iridescence, but a very short while. Usually they last only from five to ten minutes and often even not so long, although the yellow color may generally persist on the edges for as much as half an hour, more so in the morning than in the evening hours.

4. clouds of incandescent metal

These clouds are most wonderful to behold, and the intensity of the light emitted by these incandescent lumps is such that it baffles description. I have never before seen anything of this kind in the Alps or elsewhere. One can see all the nuances of color exhibited by heated metal or coal, from dull red to blinding white incandescence such as is seen in silver furnaces known in German as the "Silberblick." But most generally these clouds present the appearance of lumps of glowing coal surpassing, if anything, the latter in brilliancy and intensity of color, and the sense of sight is still more completely deceived by the gradual burning away of the glowing mass, offering to the eye the spectacle of a mass of charcoal which is being quickly consumed in a furnace with a very strong draft.

How can the intensity of the light emitted by these clouds be explained? They throw out at times a light which to the eye is as intense and blinding as that of the sun's disk itself, yet they present a surface many hundred times greater than that of the sun's disk.

dark radiations

Is it not possible that in this intense iridescence, not to say incandescence, we see, not only a phenomenon of reflection and refraction of the rays of light, but also, at least partially, a phenomenon of conversion of dark radiations of the sun, which causes in our eye the sensation of light? Or might it not be possible that the dark rays being absorbed in the mist in some way or other reduce the absorption of the light rays and render the process of reflection and refraction of the latter more economical?

I cannot recollect any experiments carried on with the object of ascertaining the influence of temperature on these processes. A hot glass lens ought to be more efficient in letting the light rays through than a cold one. But, reasoning in the same strain, it would appear that reflection from a surface ought to be impaired by heating. Furthermore I should think that it cannot be indifferent for these two processes at what temperature the body reflecting or refracting the rays is maintained, at least one must infer so from the accepted theories according to which the dark and luminous radiations merely differ in their wavelengths but are otherwise identical.

The most plausible view on the above phenomenon still seems to me the first: invisible radiations are partially converted into luminous rays or radiations, thus supplying the additional light which it is difficult to account for otherwise. It is not impossible that a phenomenon similar to fluorescence might be produced by heat rays falling upon the particles of mist, thus heightening the light effect, or they may be caused, by the dark rays, a decomposition or falling apart of the vapor particles (as Tyndall demonstrated) — and this process may be accompanied by some evolution of light. Certainly the particles capable of producing such vivid iridescence must be very minute, much smaller than ordinary particles composing the clouds, and their form cannot be but a passing one, as is evidenced by the rapid disappearance and reappearance of clouds.

These four types of cloud, which can be observed here almost daily and which in purity, brilliancy and depth of color and sharpness of outlines surpass by far such clouds noted in other parts, constitute the chief attractions of the incomparable beauty of this sky. These phenomena would be more appreciated if they were more rare, but the fact is that for most people they loose a large portion of their charm by forcing themselves upon the eye too frequently.

sunshine, x-rays

We speak of "Sunny Italy," but compared with Colorado, that country might be almost likened to foggy England. They tell me that there are scarcely ten to twenty days in a year, on the average, when the sun docs not shine, and even this estimate is rather exaggerated. Since my arrival here about the middle of May, with the exception of a few passing thunderstorms, the days were clear with just enough clouds in the sky to break the monotony of the blue.

No wonder that consumptives and generally people in feeble health are getting on here so well. The purity of the air, the altitude, which compels exercise of the lungs to be continuously and unconsciously practiced owing to the lesser density of the air and smaller percentage of oxygen (about 20 percent less than at sea level), the dryness of the air which is altogether exceptional, all these causes may cooperate more or less efficiently in improving the condition of the patients, but I believe that the chief cause of betterment is to be found in the profuse and cheering sunlight. Whether the light produces a specific germicidal effect is a matter of conjecture.

I learned here that experiments had been carried on to ascertain whether there are any Roentgen rays emitted by the sun or produced in other ways by the sun's rays, but the results were negative. Similar experiments, I am told, were conducted for a long period on Pike's Peak, but no action on a photographic film, which was the means of these investigations, was noted, at least not such as might be attributed to Roentgen rays.

consumptives

I think though that rays of this kind must be ultimately demonstrated to exist in the radiations of the Sun, as well as of most other sources of intense light and heat. It is possible that such rays are, in a measure, active in arresting the process of decay caused by the bacillus. I conclude that, since the bacillus of tuberculosis is an organism developed under exclusion of light, such rays of short wavelength, made by any means, to penetrate the tissues and reach the affected parts of the same, must needs be inimical to the development of microbes not used to such rays.

Though this conclusion might not prove true, still there is a good foundation for it, and I am hopeful that, with the apparatus I am now perfecting, it will be possible to produce Roentgen rays of great intensity which will furnish the long-sought-for means of successfully combating these dreaded diseases of the internal organs.

Whatever be the cause of the marvelous improvement noted in patients, it is a fact that most people afflicted with these ailments, and often pronounced beyond medical help, recover and soon get quite well here. A short while ago I was induced by a friend to go to a dinner he gave in my honor where I met a number of more or less interesting people. The conversation during the entire evening was an animated one and the entertainment highly enjoyable. Everybody seemed to be in high spirits and excellent health. But

my pleasure was spoiled in the end when I learned before parting, with painful astonishment, from a friend who is a very skilled and competent physician, that of the two dozen people I met scarcely one individual had more then one whole lung left, the majority of them being in fact "much farther gone" as he said, so that they would infallibly die in a very short time if they would leave here.

I soon learned that there were thousands of consumptives in the place, about the only healthful people being coachmen, and I concluded that, while this climate is certainly in a wonderful degree healthful and invigorating, only two kinds of people should come here: those who have the consumption and those who want to get it.

That the sun's light and heat exercise a highly beneficial effect on these sick people may be inferred with certainty from its effect upon people who are quite well. It is curious to note how agreeable and indispensable the sunshine becomes here after a while. Even healthful people become sad and unstrung when the sky gets clouded and dark. I have however observed such an effect before, but it is quite natural that it should be so here where the sun shines constantly day after day. I do not suppose that in London or even in New York, where the weather is comparatively fair, that much attention is paid as to whether the sky is clear or clouded, but here every laborer laments when the sun does not shine.

Despite the beautiful spectacle offered by the parting sun, one feels sad when its disk sinks behind the mountains, and one is thoroughly glad to see it rise again. These feelings are experienced, of course, everywhere, but somehow they are of greater intensity here than elsewhere. Considering the elevation, the small density and exceptional purity and extreme dryness of the air, the scantiness of the vegetation and particularly the scarcity of protecting timber, the vastness of the practically desert prairies over which the wind can sweep unimpeded, it is not difficult to guess the general nature of the weather in Colorado.

climate data

Nevertheless it is a surprise to learn that the climate is mild in an extraordinary degree. The storms come but seldom and last one or two days at the most, the snow remaining scarcely over more than thirty-six hours on the ground.

In fact Colorado people seem to be particularly proud of their winter climate. I expressed to a friend my delight at the wonderfully fine and bracing weather we had so far, but he astonished me by saying: "This is not a fair opportunity to judge. To form a correct opinion of the qualities of this climate you must come here in wintertime." I could scarcely conceive how it could be possibly finer and more agreeable then so far experienced.

I expect to get data as to the pressure, temperature, moisture, etc. The pressure at present is about 24 inches average, considerably less than at sea level, but, owing to the bracing air, one does not feel much the effect of the rarefaction of the atmosphere except when performing some physical work, when one gets quickly out of breath.

The humidity must be extremely small, otherwise one would feel both the heat and cold much more. The mean temperature presently at noon is about 80 degrees in the shade, but in sunshine it is different. I believe the good people here are more or less inclined to find the days in summer cooler than they are in reality, and they seem also to prefer to be silent about cold snaps, which occasionally come in wintertime. But from some indiscreet persons I have learned that the thermometer was at times very near 40 degrees below zero, and, in the plain sunshine of summer, it is apt to be "way up" as my informants told me. I feel sure that it can't be far from 150 degrees. The power of the sun's rays on certain days when the atmosphere is particularly calm, dry and pure, is such as to positively surpass belief.

heat hazards

The water pipe passing for some distance across the field to the laboratory being partly uncovered, the heat was as a rule so fierce that the water came out boiling and steaming like in a Russian bath. It would be impossible to hold the hand in it, even for a few moments, for it would at once cause a severe pain.

One day, about five o'clock in the afternoon, the rays fell through the open door on a high tension transformer which I had brought from New York, and, before anybody could notice it, melted out all the insulation, rendering the apparatus completely useless. I observed the danger a few days before and warned the assistants to watch the machine, but unfortunately on that day the usual precautions were omitted.

Several barrels filled with concentrated salt solution were placed outside of the laboratory, and the pressure in them rose every day as in a steam boiler, and a few of them were damaged. When the cock was opened the water squirted out to a great distance across the field and it was thought advisable in order to avoid bursting and damage, to leave a small opening in the barrels for the escape of the steam.

The most astonishing experience of this kind was, however, the heating of a wooden ball covered with tinfoil, which was supported above the roof, to a point that it was deemed unsafe to expose it to the sun's rays. It emitted a dense vapor actually like smoke, and the tinfoil crumbled away! This excessive heating seemed to take place suddenly. I believe that it occurs when, owing to the removal of a layer of impure air, a particularly clear path is opened for the sun's rays, which then pass through the pure medium without much loss. Often I have felt a scorching pain on the cheek or neck to come on suddenly when working in sunshine, and I can only explain it with the above assumption. But the most interesting of all are the electrical observations which will be described presently.

AUGUST 28

On this day, Tesla is experimenting with various radio-receiving apparatus, including a "sensitive device" but also lightning arrestors (series spark gaps), these being stimulated by his magnifying transmitter and also by natural lightning. (Colorado Springs is probably the top lightning hot spot in the USA.) Even when lightning strokes are no longer seen or heard, Tesla observes evidence of them in his receivers. Eventually he turns his attention back to the sky and the mysterious phenomena of its electrical display:

… The rain and lightning were just beginning. Magnificent intense white light witnessed below Pike's Peak, something very unusual. It resembled a white-hot silver furnace. The lightning on the mountains was very frequent, and the discharges of unusual brilliancy. Twice a curious phenomenon was noted. Lightning striking in one part of the mountains from cloud to earth was seen in another part, a few miles away from a high peak, passing from the peak to the cloud. The discharge was much thicker at the root and branched out towards the sky, spattering itself in many branches and disappearing in fine streams. The astonishing phenomenon was witnessed a second time and subsequently, and a similar discharge took place from other peaks.

Is it possible for a discharge to go from Earth to cloud? As far as the visual impression is concerned, there can be no doubt. The discharge in all cases followed a preceding lightning discharge in another region, and apparently from cloud to earth. Perhaps it can be the effect of an intense vibration started by the first discharge, which results in another discharge towards an oppositely charged cloud.

The clouds were unusual in configuration and grouping. A large portion of the sky was quite clear. The wind at times was very strong. An instrument by its constant play indicated strong electrical disturbances through the earth, even when there was no display of lightning as far as could be seen or heard.

gimme a break

Appendix D

The Tesla Mystique

by George Trinkaus

DESPITE HIS OBSCURITY, THE GREATEST GENIUS OF ALL TIME was Nikola Tesla. Geniuses like Tesla and Einstein come along only every 50 years or so. Tesla was a humanitarian idealist consumed by a passion to save the world from poverty and war. An extraterrestrial from Venus, Tesla was a superhuman inventor who had the uncanny ability to visualize the operation of machines in his head.

Tesla, the prodigal genius, the forgotten genius, the sorcerer, the man out of time, the wizard with lightning in his hands. Tesla's abstruse, esoteric electric technology is beyond the average intellect and is best interpreted by qualified, degreed experts in quantum physics and scaler electromagnetics.

In his old age, forgotten and penniless, Tesla was murdered by agents of the U.S. government in a seedy Manhattan hotel room, his papers confiscated and disappeared by the FBI. Tesla's technology continues to be undisclosed to the public and instead is directed into black projects, like the Philadelphia Experiment, and HAARP.

Antigravitic UFOs are Tesla technology, as are the quantum-vacuum, zero-point energy generators that drive them. Tesla's technology has brought us such products as the Tesla Scaler Potentizer, the Tesla wristwatch, the Tesla Teletransporter, and the Tesla Male-Enhancement Helix, as well as the Tesla electric sportscar. It was Tesla's magnifying transmitter that caused the devastating Tunguska explosion of 1908. A contemporary black-project version underground in Canada accidentally brought about the East Coast blackout of 2003.

The preceding will pass as intelligent Tesla-speak on late-night talk-radio, on the Internet, and at cocktail parties. The preceding contains some direct quotations from the literature; other examples are more periphrastic; all are consistent with the idiom of The Tesla Mystique.

The Mystique sentimentalizes, romanticizes, and mystifies the memory of Nikola Tesla. The Mystique is determined to make Tesla awesome, fantastic, and beyond comprehension. The Mystique is an intellectual fashion wave that distorts biography, history, science, and consciousness. It is a vogue that has been embraced by fringe media and dabbled with by mass media, where Tesla is at risk of being packaged as cliché. The Mystique is taking root in the culture as official truth. The Mystique has no critics.

The Mystique must make Tesla a paragon of character in all directions. For example:

1. Tesla, the Humanitarian

The sentimentalized Tesla becomes the humanitarian-idealist scientist fighting the establishment for the benefit of human society. But contrary to the Mystique, Tesla was above all an engineer. Engineering was his education, his consuming passion, his daily practice. Tesla the engineer incidentally might have hoped that his inventions would advance society toward peace, convenience, comfort, and a more efficient use of human energy — incidentally.

Tesla was indeed an idealist. His idealistic passion was that of the engineer. He insisted that machine possibilities be carried to their ultimate, logical, evolutionary conclusions. He had an incorrigible respect for his own inventive instincts. Let's respect Tesla for all of that in his character. Is it not enough? But the Mystique requires a saintly paragon. The Mystique would give us a liberal humanitarian Tesla, politically correct according to the modern etiquette.

Tesla's inventing pushed forward relentlessly, sometimes oblivious to ruling-system interests. That is heroic. Tesla's dream for the ultimate realization of his wireless was a "World System." His ill-fated tower at Wardencliff was to be the prototype magnifying transmitter for a global communications network. It was to propagate both broadcast and point-to-point wireless, including telephone, telegraph, stock tickers, teletype, even FAX, as well as voice and music on a global scale, a corporate monopolist's dream.

The Wardencliff magnifying transmitter also may have had the potential to be a utility that could propagate electric power wirelessly through the earth to industry and homes in the New York area and perhaps beyond. But this potential is not cited in the "World System" brochure that was published to promote the project. Did Tesla advertise this wireless-power capability to his financier, J.P. Morgan? Did a humanitarian Tesla threaten capitalism with free electric power and thus drive Morgan to halt the project, as the Mystique would have us believe? Not necessarily. The project's formidable world telecommunications capability in itself may have been enough to induce Morgan to kill Wardencliff.

Tesla's grandiose dream would have taken radio right off the bat into a global centralization not quite achievable even today. No multinational institutional structure, corporate or governmental, existed at the time upon which such a system could have been

Wardencliff

founded. At the turn of the century, the J.P. Morgans might have been dreaming globally, but it would be thirty years before the moguls could establish radio networks on just a national scale.

The Mystique would give us a liberal-humanitarian Tesla, politically correct according to the modern etiquette. Unfortunately, he does not fit the template. Pure technologists like Tesla tend toward a mechanistic social view. To Tesla, human society was a machine, and it needed perfecting. Tesla saw his World System as a civilizing force. He wrote, "It will be very efficient in enlightening the masses, particularly in uncivilized countries and less accessible regions."

What did Tesla mean by "civilization?" He said, "No community can exist and prosper without rigid discipline." He said, "Law and order absolutely require the maintenance of organized force." Tesla, a believer in organized military force, invented weapons of war (robotic boats and submarines, death rays, etc.) which he tried to sell to the US Navy and to the Department of War.

The Mystique would make Tesla a pacifist, because the inventor idealized a mechanized, automated, robotic warfare which would replace human participants entirely with machines (a misunderstanding of the institution of war, which depends upon human destruction and terror). Tesla foresaw weapons of mass destruction and the possibility of a lasting world peace based on mutually assured destruction, but this is a Kissinger pacifism.

Tesla, the inventor of wireless and remote control, foresaw the "teleautomatic" warfare of today in which robotic Predator aircraft, controlled from a bunker in Nevada, deliver bombs and missiles upon Afghanistan. Sorry, but if we credit Tesla with remote control, this invidious connection can be made.

Tesla, the purist engineer, advocated a social engineering that included eugenic cleansing. He said government "should prevent the breeding of the unfit by sterilization and the deliberate guidance of the mating instinct."

Humanitarian or totalitarian? You decide.

2. Tesla, the Genius

Granted, if the word genius has any meaning, it would apply to a Nikola Tesla. The issue is, how much meaning can any word have when it becomes cliché? When you utter "Tesla" in conversation, if you don't hear "who?" you will get back "genius" within five seconds. Try it.

Tesla could visualize machines working in his head! But can't any mechanical thinker? Yet one hears this all the time.

Is there a Tesla biography with a title that has in it no "genius" or variation thereof? A magazine journalist of the 1940s, John O'Neill, authored a panegyric called *Prodigal Genius* (1946), which has become institutionalized as the standard. Biographer O'Neill's enthusiasm may have been genuine, his biography eloquent and respectably researched, but his spin is an echo of the newspaper hype of Tesla's heyday. O'Neill's biography (still reprinted today) set precedent for the obligatory promotional idiom that permeates almost all of Tesla biography as well as almost all other discourse on the man and his work.

"Tesla, the great mathematical and physics genius, came up with an idea called zero-point energy," quacks Michio Kaku, the string theorist, who has been declared a genius himself.

"Would the electrical age never have happened without Nikola Tesla? Perish the thought, says the Mystique."

In his heyday, Tesla was exploited as the poster-boy for the emerging electric-power utility industry that was exploiting his AC inventions. Far from obscure, Tesla circa 1900 was as famous as Thomas Edison. The press romanced a genius-Tesla to the public in the process of promoting this industry, which would develop into the omnipotent monopolies of Samuel Insull, the modern grid, and Enron. Of course, later, the media would turn on their pet genius and try to render him invisible, limiting his exposure to an annual birthday press conference.

The Tesla brand was used by the system and then abruptly disposed of, but it has seen a little revival today in the brand names cited in the introductory satire, which you can Google (except for the Male-Enhancement Helix, which I made up.) The Tesla sportscar does have a plausible claim in that today's electric cars use AC motors derived from Tesla's.

Would the modern electrical age never have happened without Nikola Tesla? Perish the thought, says The Mystique. Tesla's alternating-current meshed with the needs of an industrial system which could not have expanded nationally into today's expansive grid on Edison's puny direct-current system. But it's arguable that the alternators, motors, and transformers which the system needed in order to progress might have been invented by a "genius" other than Tesla's. The Mystique shudders at the thought.

Tesla inherited a pre-modernist physics that allowed him exceptional latitude for exploring technological possibility, which may be one reason modernists, in their envy, feel they must ostracize him from humanity as a special case, a genius. Some differentiate Tesla as so exceptional that he must have come from another planet.

The mad-scientist cliché is another instance of the differentiated, special-case Tesla. (That genius Tesla was so weird, possibly insane.) The scary spark streamers in horror films issue from the coils of that mad scientist, Tesla. Biographers dwell on the eccentricities of their genius: the compulsive hand-washing, the stack of napkins at Delmonico's, the refusal to shake hands. Wrapping Tesla in cliché is one way not to see him.

Eccentric genius cliché aside, we can responsibly describe Tesla as having exceptional insight, that he was a sensitive, meaning one who has retained the original insight of the child, that he was a visionary. This is certainly supportable, especially when argued from the borderland by a Gerry Vassalatos.

One work that provides relief from the pervasive genius cliché (although the title does have that ring) is *Enigma Fantastique* by W. Gordon Allen (Health Research). The book parallels Tesla's life with that of Rudolf Steiner. It was Tesla's distinctive education that made him special, says Allen. Jesuit instructors played a part, as did various mystical schools circulating in Eastern Europe when Tesla studied at Graz. These stressed an unusual yoga-like self-discipline of both mind and body and the development of powers of rigorous self-application. Tesla's distinctive strengths are these. The genius sentimentality is a disfavor to Tesla and to the concept of human potential in general.

3. Tesla, the Victim

The victim cliché dovetails with the genius one, and it feeds upon inventors (as it does writers, musicians, and artists). Thus the mystique would have our genius dying in poverty. His death by murder is sometimes in the script (by government agents), and the myth has all of his work stolen and suppressed.

It's true that, after being dumped by J.P. Morgan, Tesla suffered economic humiliations. For example, there is evidence that he was forced to pawn his interest in Wardencliff to the Waldorf Hotel in an attempt to cover his debt there. But the Mystique fails to appreciate that Tesla died at the ripe old age of 87, of natural causes, not in poverty, but in his rooms at the New Yorker, a commodious midtown hotel, which is not quite the Waldorf but a nice situation for a senior citizen, being a little city unto itself. The New Yorker back then had 2000 rooms, five restaurants, various shops and services, and even its own power plant (DC) and a radio station (carrier-current). It was a nifty habitat, testifies this writer, who stayed at the New Yorker more than once in his college days, fifteen years after Tesla's death. The New Yorker Hotel was a very decent place for a venerable inventor to live out his last days. Culturally the cosmopolitan Tesla had become a New Yorker.

Tesla's papers and belongings at the Hotel New Yorker indeed were confiscated by the government, not by the FBI, but by a department that once existed within the Department of Immigration called the Division of Alien Property. It may be true that many of Tesla's New York notes got disappeared, and we may yearn to see them, but it is a distortion to dwell lugubriously on this, for we have so much of Tesla's technology raw – in a hundred or so US patents in print for years and now easily accessed on the Internet. Also, printed in hardcover volumes have been Tesla's collected *Lectures, Patents, Articles* and a rich and copious document on radio technology called *Colorado Springs Notes*. Tesla, a gifted writer, wrote his own little autobiography called *My Inventions*. This writer has edited and published Tesla's *The True Wireless*. (See Appendix C.)

Consider also all the material unearthed by Leland Anderson, John Ratzlaff, Gerry Vassilatos and others. Vassilatos mined Tesla's notes in archives found in the annex of the New York Central Library. Biographer Marc Seiffer mined the files of the National Archives in Washington, DC, and in the Tesla Museum in Belgrade. By no means has the available wealth of Tesla material been totally explored and digested. Those who are

demanding government "disclosure" of free energy and antigravitics: have they exhausted the government's open patent archives? Also a lot of accessible Tesla has been ignored because it is just too nonconforming and deep.

Many accounts tell of Tesla being the victim, not just of J.P. Morgan, but also of George Westinghouse. It's true that Westinghouse signed that dollar-per-horsepower contract with Tesla from which he might have made millions in royalties on the alternators, motors, and transformers that Westinghouse manufactured. It is true that Westinghouse tore up that generous contract. But it is generally unappreciated that J.P. Morgan was standing behind Westinghouse with a pistol to his back, so to speak, for Morgan then was financing Westinghouse, as he financed Tesla, Edison, and other US industrial pioneers. J.P. Morgan was the banking link between New York and the London House of Morgan. Such pipelines of capital from Europe drove the US industrial revolution.

It is true that Tesla's later work invited suppression, but so many of his inventions did make it into patent. The system tried to delete the Tesla name, but it nevertheless persistently did live on, underground, floating on the inventor's former fame, until resurrected in the 1980s, albeit unofficially, and Tesla's fame has been growing ever since.

Victim? How many who have ventured into the uncertain profession of inventing can claim such good fortune? Given the fate of inventors generally, Tesla, the victim, fared very well. Compare the innumerable inventors whose work goes forever unknown, including so many who did make it into patent, whom you might run across in a subject search, find just as clever as that genius Nikola Tesla, but who will remain forever in obscurity never to be celebrated by any mystique.

4. The Philadelphia Experiment and Other Folklore

Did Tesla have anything to do with an item of folklore called "The Philadelphia Experiment?" This sci-fi fantasy, widely circulated as serious history, tells of a Navy experiment in making warships invisible. The experiment went haywire, goes the story, dematerializing a vessel in Philadelphia only to have it rematerialize, crew and all, in Norfolk. Some crew get stuck in the bulkheads during the Einsteinian space-time transition. The Philadelphia Experiment has become a major fixture in the Tesla Mystique. "Tesla technology" is vaguely imputed to the phenomenon, and the story (evidently a tavern yarn that got into book and out of control) has Tesla on board as a technical participant.

The story is set in the year 1944. However, Tesla died in 1943 (January). A review by the Department of Naval Research found no evidence to support the tale. An independent military historian searched in vain in all imaginable Navel records for any clue that would corroborate this incredible event and published this research in the alternative *Electric Spacecraft Journal* (No. 8).

Does Tesla's peculiar radio technology really have anything to do with another black project called **HAARP**? This gigantic radio-transmitting complex indeed exists in Alaska, and is said to be situated over a huge reservoir of natural gas used to fuel its megawatt dynamos. What a power supply for any transmitter! Very ominous. But is it Tesla?

Indeed HAARP is radio, which Tesla invented. But Tesla's radio is longwave, and HAARP's carrier is shortwave (or VHF, depending on your source). These upper bands (3 to 300 mc) Tesla may have never explored, for he knew that they were relatively ineffective compared to the low frequencies (under 500 kc). Of course, HAARP could pulse its carrier to any beat it wants. Also HAARP's alleged experiment is said to involve the stimulating and heating and bouncing of energy off of an ionosphere, which Tesla insists does not exist and would have absolutely no effect on radio propagation if it did (*The True Wireless*).

Finally, I'm guessing that the oscillator technology at HAARP is a conventional feedback system using vacuum tubes or transistors, the standard since the 1920's. (If you know otherwise, please advise.) Tesla's powerful oscillator, however, is a high-voltage, sudden-pulse disruptive-discharge system (Radio Tesla).

Can a giant transmitter create hurricanes and other destructive storms? HAARP could be a weather-mod weapon. If you charge the atmosphere with electricity, you can exert pressure on a system. That is, I've seen equations translating electric charge in columbs into air pressure in psi (Yost). HAARP may project electric power from its abundant supply up to a weather-mod satellite equipped to beam power down to the target on a conductive ultra-violet beam, a more likely scenario than targeting by "ionispheric bounce," a theory Tesla detested. I've heard that just such a satellite was put into orbit in 2005.

Earthquakes? Can a radio transmitter's power be similarly directed to a geographic target and execute mechanical effects upon tectonic plates? Could there be perhaps some kind of piezo-electric effect? I'm waiting for the data. Inform me if you know. Tesla did invent the equipment to do the job. His "earthquake machine," or tele-geodynamic oscillator, was not a wireless transmitter but a mechanical device, an acoustic vibrator that worked on the mechanical plane, and it could create seismic effects at a distance.

That a Siberian forest near **Tungusta** was leveled in 1908 by a 15-kiloton blast from Tesla's magnifying transmitter at the opposite side of the earth is a bizarre rumor that will not die. Never mind that Tesla at this time had no magnifying transmitter to play with, assuming this machine could conceivably have any such power.

That a super-size Tesla coil secreted underground in Canada accidentally created the summer-2003 East Coast electric-grid blackout is a story that played on Art Bell one night but never gained traction.

The Patents may be the only sound footing in the mythic world of Tesla.

UFOs, antigravity? Both are associated with Tesla. But there are just a few speculative notes, and his few aircraft patents use conventional air-flow lift. Some antigravity experimenters employ the Tesla-coil as a high-voltage power supply. Some UFO researchers speculate that flying saucers use dual out-of-phase Tesla coils in the levitation drive.

Compounding the Tesla Mystique is the inventor's anticipation of technologies unknown to the general public until the 1960s, like fluidics, cryogenics, and computer logic. That Tesla is original and has priority in these arts is assumed by enthusiasts, but confirmation would require extensive searches among patents filed by other inventors of the period. So much innovation gets lost in the patent archives.

Tesla is celebrated for an interesting electric-ray device which he introduced to the press, unfortunately, as a **death ray**. He explored this clever vacuum invention in laboratory prototypes, and drawings exist in the literature, but he never put it into patent. Because of one cavalier utterance by Tesla at a birthday press conference, the death ray was sensationalized by the media and given disproportionate prominence compared to many other newsworthy Tesla inventions that did get into patent but which the press ignored. All of Tesla's other fascinating work in electric-ray technology is generally ignored and unappreciated even today.

Did Tesla build a **free-energy Pierce Arrow**, or is this story more of the Mystique? A writer purporting to be the inventor's nephew tells of a 1931 drive in Uncle Nikola's customized Pierce Arrow. Powered by an 80 hp electric motor supplied by a vacuum-tube space-energy receiver, the car achieved 90 mph. Now this would be the ultimate electric vehicle, but Tesla's real nephew has come forward to debunk the story.

So much fantasy, folklore, and disinformation swim around Tesla that a writer venturing into these waters has difficulty finding any secure footing. This writer had the good fortune of first encountering Tesla solely in patents (having stumbled upon a complete set that someone had photocopied at the National Archives, prior to these becoming available in book). Surveying the field, even as the literature was developed back then in the early eighties, this writer soon concluded that the patents may be the only sound footing in the slippery mythic world of Tesla.

It's a good idea for any writer on Tesla to ground himself in some hands-on experimental Tesla circuit-building projects, such as the Tesla coil, before he claims to know his subject. Such grounding can help protect a writer from being consumed by The Mystique. To focus on Tesla's life instead of on his work

opens a door into The Mystique, and a special objectivity and discipline is required. It may be safer to start with the technology and work outward. Almost all of contemporary biography has been hopelessly infected by the Mystique.

5. What Tesla Really Did

These are the contributions of Tesla to the technology of civilization, to the patent archives, to knowledge:

Tesla invented the 60-cycle AC power system that runs civilization today, the dynamos, transformers, motors, regulators and arc lamps. This technology established, along with his own wealth and fame, Tesla went on inventing. A turning point was 1891, when Tesla applied for patent 462,418, a Method and Apparatus for Electrical Conversion and Distribution. His first high-frequency lighting patent, the system was powered by a spark-gap oscillator like that which would drive his Tesla coil. In 1891 began a stream of inventing that produced the Tesla coil, radical non-filament lighting devices, the x-ray and other vacuum electric-ray devices, electrotherapy, the magnifying transmitter, radio, wireless power, and even a space-energy receiver.

Although much of Tesla's high-frequency work did get into patent, most never went into manufacture, and some which did, like radio, did so under another's name, like Marconi's. (Tesla's priority in radio, after decades of litigation, was finally decided by the US Supreme Court in 1944.)

Tesla never gets credit for inventing two fundamental radio-receiver circuits, the regenerative and the superhetrodyne, but schematics for both can be found in Tesla's Colorado Springs Notes of 1899.

Few of Tesla's explorations into ray technology got into patent, and this third phase of his experimental life is under-documented considering its implications. (See the work of Gerry Vassalatos published by Borderland Sciences.)

Tesla's past includes a misty period in Canada in the 1930's which may have included accomplishments unsung. A reader in rural Quebec, a radio amateur, wrote to me that locals talk of Tesla constructing a practical wireless utility that successfully transmitted power for 75 miles from Chambord to LaTuque.

Free energy? Tesla did patent a fundamental free-energy concept in 1901 (Patent 685,957). A space-energy receiver collects ambient energy and converts it to a practical electric output. A very crude space-energy receiver is the solar-electric panel. That Tesla pursued space-energy into any working prototypes is very likely but difficult to document. Space energy is a truth evident in a lightning bolt, but it is a most taboo truth.

Lurking perhaps in Tesla's confiscated notes are drawings of practical table-top space-energy power plants. Subsequent inventors have demonstrated similar successful devices (Morey, Plauson, Coler, Hendershot, Stubblefield...) but they have received more punishment than reward for their efforts.

Tesla said, "Electric power is everywhere present in unlimited quantities and can drive the world's machinery without the need for coal, oil, gas, or any other fuel." This truth is a most taboo truth, but it is evident in a lightning bolt.

The Tesla coil is arguably a free-energy device, or at least an energy magnifier. Tesla said his coil "takes on a 110-volt direct current circuit, according to load, and adjustment, from 5 to 30 watts. It gives a powerful stream of sparks 6 inches in length, but if desired this distance can be easily doubled without increasing the energy consumed; in fact, I have found it practicable to produce by the use of this principle sparks of one foot in length involving no greater expenditure of energy than 10 watts." (Lecture before the New York Academy of Sciences, April 6, 1897, pg 43, Leland Anderson, ed.)

It's fashionable to say that the conversion technology required to put free energy into practice is way off in the future or locked up in government files awaiting "disclosure." Yes, but only if you ignore all of the archived patents and other available literature, which in the aggregate would supply more than enough knowledge for engineers to proceed with confidence to the developmental workbench.

If it is true, as Tesla said, that electric energy is everywhere present and can be harnessed for practical use, then energy-scarcity must be all myth. Scarcity, one could argue, is contrived and advertised, not on the basis of any known scientific truth, but upon scientific fictions propagated by well-paid researchers and academics in order to better control the population. The latest wrinkle in this ongoing propaganda campaign is "peak oil." The entire world has been vaccinated with a conviction that electric energy is a finite, limited, esoteric resource that must be centrally generated by obnoxious combustive and nuclear processes or by those ridiculous, unsightly windmill farms. Electricity is sold as a scarce resource that must be conserved, controlled, and paid for by the kilowatt hour.

Contrived scarcity is the ruling system's rule in respect to all energy sources, electric, petroleum, or whatever. Tesla ran afoul of this rule of rule.

The literary contributions of Tesla-the-genius are largely unsung. As a writer, I particularly appreciate the eloquence of Tesla, manifest even in the language of his patents, which I enjoyed quoting liberally in my Lost Inventions, etc. Poetic eloquence was not unfashionable in premodernist science. Tesla was quick to pick up a pen and left us a prolific literary legacy of elegant articles, lectures, and patents. His notes are no exception, including the published *Colorado Springs Notes.* (A good example of Tesla the writer in a nontechnical vein, is his description of the 1899 Colorado Springs environment in his entry for August 1.) Tesla, a multilingual cosmopolitan and a friend of Mark Twain, exhibited an unabashed literary flair in his English, and also in some European tongues. Multilingual-cosmopolitans with a literary flair, like Tesla, are a European product, rarely bred in America and most comfortable in the USA in Manhattan.

6. The Quantum Tesla

It was inevitable that quantum, which is buzz-word number one in fashionable scientific parlance, would attach itself to buzz-word number two, Tesla. Quantum true-believers are uneasy with Tesla, and think it necessary to invest a huge volume of verbiage and intellectual energy in a misguided effort to reconcile Tesla's electrics with modern quantum theory. Tesla's science was premodern, unfortunately, so fashionable quantums must "interpret" Tesla in "correct" modernist terms.

"There is no such thing as an electron."

In a press conference in the 1920s, a reporter asked Tesla to comment on the great achievements of the electron. Said Tesla, "There is no such thing as an electron." That item of quantum vocabulary "electron" is hard to find anywhere in his writings. Somebody should print this quotation on tee-shirts and conduct a campaign to stamp out the fashion that requires such excruciating utterances as: "A Tesla coil is a quantum action device … if the phase of a split quantum particle is changed, its conjugate partner instantly knows …"Tesla took for granted a science that had served research faithfully for at least 150 years. This Victorian inventor (who was "a man out of time," according to the Mystique) had no respect for the quantum theory that was becoming fashionable in his own lifetime.

Like quantum, Tesla had no good words for Einstein's hallowed theory of relativity, which he called "a mass of errors and deceptive ideas violently opposed to the teachings of great men of science of the past and even to common sense." Einstein "wraps all these errors and fallacies in magnificent mathematical garb which fascinates, dazzles and makes people blind to the underlying errors." The theory's exponents, he said, are "metaphysicians, not scientists." He said "Not a single one of the relativity propositions has been proved."

"He described relativity as a beggar wrapped in purple whom ignorant people take for a king. In support of his statement he cited a number of experiments he had conducted as far back as 1896 on the cosmic ray. He has measured cosmic-ray velocities from Antaurus, he said, which he found to be fifty times greater than the speed of light, thus demolishing, he contended, one of the basic pillars of relativity, according to which there can be no speed greater than that of light." — *New York Times*, 7/11/35.

Nevertheless, quantum interpreters like to mix in some Einstein with their Tesla. From the same Einsteinian fashion that gives us time-warps and warp-speed, comes the likes of: "One way to visualize a Tesla scaler wave is to regard it a pure oscillation of time itself." What a turn-off is this jargon to the innocent who encounters it in a search for some real knowledge of Tesla technology!

Go all the way with the Einsteinian-quantum-electron and you'll eventually get to the glib, buzzy Dr. Kaku, who has his electrons "darting in and out of parallel universes." (Admit this language into your mind as meaning, and it is your loss.)

Tesla was not a theoretical physicist. Although his writings include some very articulate and suggestive musings in both physics and metaphysics, he asserted no systematic dogma. This is unfortunate, because it left a vacuum to be filled later by quantum-educated Tesla enthusiasts who take Einstein, Star Trek, and Kaku for granted. It becomes another way of suppressing Tesla, by co-option, by obfuscation.

"A tenuity beyond perception."

Tesla thought science took a wrong turn when it adopted the quantum and Einsteinian. Premodern science is qualitative. Modern science is quantitative. It materializes the immaterial universe, particle-izes it, counts it, mathematizes it, while discounting the qualitative, the immaterial, the subtle, the etheric, the cosmic, the poetic. You may be quantum and Einsteinian, but it is an error to impute any of this constrictive dogma to the expansive, visionary Nikola Tesla.

Tesla came in on the ground floor of experimental particle physics with his exploration of x-rays (co-incident with Roentgen's explorations). X-rays, he once observed, were "streams of matter." The vocabulary of particle physics he otherwise avoids. The materialistic philosophy of quantum, which promises to find the universe in a quark, would be completely alien to Tesla's cosmic view, in which the ether is the universal continuum, immaterial, a field permeating all of matter and all of space, "a tenuity beyond perception."

A unified field? No problem. Tesla was not straining for a "unified field." He inherited the concept. The pervasive universal etheric continuum is the medium of all electric phenomena. Charge, potential, polarity, conduction: all can be understood as instances of some stress, or disequilibrium, in the continuum of the ether. Free energy for Tesla would be derived from a disturbed spontaneously energetic continuum, the taboo ether.

Ether theory is unbound by the sacrosanct speed-of-light C-constant of Plank and Einstein, which is used to preclude the possibility of instant action. Ether theory allows for phenomena at "superluminal" velocity. That is, a disturbance of the ether's equilibrium at one point can create an instantaneous corresponding disturbance elsewhere, like the action of a mechanical lever.

Quantum theory fragmented the continuum into particles, but physics had to call in the clever Dr. Kaku to tie it all back together with "strings."

A tenuity beyond perception? Way too tenuous, say the modernists, who must materialize this mystery into "a quantum

radiometer and quantum box

my quantum radiometer

A reader sent to me as a gift a radiometer. (Thank you, Craig.) Tesla called the Crookes radiometer the "most beautiful invention ever created" and a new step in motor technology, "the jewel of motors."

My gift radiometer is a nice little toy from Tedco, but the box copy, written by a quantum true-believer, attributes the rotation to an atomic heat phenomenon. The dark side of the vanes somehow get more warmth from the light, thus reacting with "freely moving particles of air," which are assumed to be residual in the vacuum and entirely responsible for the motion imparted. Thus, "when the atoms strike the dark vanes, they kick away at terrific speed," smugly declares the box copy.

So if this is a heat phenomenon, why do the vanes turn in the cold light of my LED flashlight? And why do they rotate when I take my CB transceiver hand-unit, hold the rubber-ducky aerial to the bulb, and key up? Also the bulb fills with a mysterious milky luminescence, reminding me of Crookes' high-voltage vacuum-bulb experiments.

The quantum mind cannot handle a dynamic vacuum and must throw in a few residual atomic particles to get some action. Leave it to the quantum believer to destroy a powerful mystery with some stupid materialistic explanation. Also, the box copy says simply "radiometer," giving no credit anywhere to Sir William Crookes, whose rich premodernist work has apparently been forgotten by quantum true-believers.

As an antidote, I dig up a 1996 article about the radiometer from *The Journal of Borderland Sciences* (now lamentably defunct) which attributes rotation to an energetic component of light and of electric rays that is black radiance. Crookes observed the phenomenon in the mysterious dark spaces that appeared in his electrified tubes and also in the spark streamers from Tesla's coils. The quantum believer must materialize dark radiance as "dark matter" or "black holes." The radiometer, like the electrified Crookes tube, "was designed to peer into astral space beyond the inertial walls," says the Borderlands writer, who reports that Crookes built a huge demonstration radiometer, electrified it with Tesla currents, and inside one could actually see the dark space pushing the vanes around, until speed resolved the rotor into a whirling darkness.

Etheric science is expansive, cosmic, metaphysical, spiritual, and way more fun than quantum, which is myopic, submicroscopic, materialistic, atheistic, prosaic, and dull.

The challenge to the Tesla biographer is to transcend the great fog of fashionable folklore obscuring his subject. The challenge to the Tesla physicist is to transcend fashionable theory and to submerge himself in Tesla's physics, which, fashionable or not, is an occult ether physics. To comprehend Tesla one must dare to cross over into the fringe.

sea of point particles in motion." This from Moray King, who located his particle motion at a "zero point," hence "zero-point energy," or ZPE. It's amazing how this zippy vocabulary has caught on. Just don't ask its users to define their terms. Some closet etherists will cloak their theoretical utterances in quantum-speak in order to pass.

Consider the following musing from Tesla regarding the mysterious properties of the human eye: "A single ray of light from a distant star falling upon the eye of a tyrant in bygone times, may have changed the destiny of nations, may have transformed the surface of the globe, so intricate, so inconceivably complex are the processes of nature." Such an utterance from a proper modern academic is inconceivable. Horrors; it's almost astrology. However, to the premodernist mind, such reflections were part of science.

Radio theory? Quantum-style interpreters like to mix Tesla with the modernist radio theory of Heinrich Hertz, a theory which Tesla vehemently rejected. To Tesla, wireless was not a light-like radiation that is reflective and refractive and bounces off an ionosphere. Wireless Tesla understood as "compressions and rarefactions of the ether," an ether disturbance, "like a wave … in the infinite ocean of the medium which pervades all," he said.

Schuman cavity? Neo-Hertzian Tesla enthusiasts claim that, at Colorado Springs, Tesla's signals propagated by bouncing about inside of a resonant "Schuman cavity," presumed to exist between the earth and an ionosphere. What ionosphere? Tesla declared there was no such thing, and, if such existed, it would have no effect on wireless propagation. Tesla believed the earth itself was a giant capacity that could be resonated, like the terminal capacity on his Tesla coils.

Hertzians and quantumists should at least apologize to Tesla when they step all over his premodernist science in an attempt to reform it in their own dull terms.

The Tesla Network

events

The Philadelphia Tesla Fest
July
484-955-0545
www.TeslaScienceFoundation.org

Teslamania, Toronto
July
416-521-7462
teslamania.tv

Extraordinary Technology Conference
Albuquerque, NM / July-August
296 E. Donna Dr., Queen Valley, AZ 85218
(520) 463-1994

places

The Tesla Science Center at Wardencliff on Long Island
POB 552, Shoreham, NY 11786
info@teslascincecenter.org

The Tesla Museum
Belgrade, Serbia
51 Krunska, Beograd, Serbia
www.tesla-museum.org

parts

All Electronics
Surplus parts, variacs
14928 Oxnard, Van Nuys, CA 91411
(818) 904-0524
allelectronics.com

Antique Electronic Supply
Crystal detectors, tubes, variable caps and more.
6221 S. Maple, Tempe, AZ 85283
(480)820-5411
tubesandmore.com

Fair Radio Sales
Military electronic surplus, high-voltage caps, transmitters, many surprises, print catalog.
2395 St Johns Rd, Lima, OH 45804
419-223-2196
www.fairradio.com

Information Unlimited
Tesla coils, assembled and kits, high-voltage caps, transformers, other high-voltage devices, lasers.
www.amazing1.com

RF Parts
Transmitter parts, tubes, transistors, high-voltage caps, transmitter variable caps.
(800) 737-2787
435 S. Pacific St., San Marcos, CA 92078
rfparts.com

kVA Effects
Theatrical Tesla coils
(832)257-6752
teslacoil.com

associations, websites

Nexus
www.nexusmagazine.com
A multinational, multilingual print mag from Australia.
Alternative news, science, technology, conspiracy, Tesla.

Rex Research
rexresearch.com
Distributes literature about suppressed science, including Trees as Antennas, underwater, EMP.
PO Box 19250, Jean, NV 89019

Tesla Memorial Society of New York
www.teslasociety.com
(713) 417-5102

Space Energy Assoc. (SEA)
quarterly journal
PO Box 1136
Clearwater, FL
33757-1136
(954) 749-1136

Tesla Engine Builders Asso. (TEBA)
TeslaEngine.org
5464 N. Port Washington Rd, #293
Milwaukee, WI 53217
teba@execpc.com

some fundamental literature

Complete Patents of Tesla (The U.S. patents)
John Ratzlaff, ed.
21st Century Books.
Patents can be ordered individually by number from the U.S. Patent Office, Washington, DC 20231 or at USPTO.gov. On line, search for Jim Biederrich's Nikola Tesla Patent Collection.

Colorado Springs Notes
The documentation of Tesla's researches in 1899-1900 on radio and wireless power, published by the Tesla Museum, Belgrade under the imprint, NoLit. A paperback reprint is published by Barnes & Noble. Hardcover from Omni. (*Serbo Croatian Diary Comparisons*, by John Ratzlaff points out some curious discrepancies between the Serbo-Croation *Colorado Springs Notes* and the English translation.)

Inventions, Researches, Writings of Tesla
by Thomas Commerford Martin, 1894
Barnes & Noble paperback.

A Tesla Bibliography
by John Ratzlaff, 1979
From 21st Century Books, which maintains the most comprehensive library of Tesla literature.
tfcbooks.com

Nikola Tesla: Lectures, Patents, Articles (Tesla Museum) Out of print. Even this 1000-page compendium does not include all of Tesla's prodigious output.

The Radio Amateur's Handbook
ARRL (make sure you get the ARRL; there are other "handbooks.") Principles, methods, construction. The prose are muddy, overly technical and mathematical. The propagation theory is Hertzian, but this is the traditional source for radio knowledge.

biographies

Tesla, Man out of Time
by Margarete Cheny, 1981
The most widely distributed, hence most widely read, Tesla biography of our time. Originally published by Prentice-Hall, now by Simon & Schuster. Reads like a masters thesis, only enough technology to make you curious, but covers the ground.

The Secret of Nikola Tesla
Film
If there must be a Hollywoodization of Tesla, let this juicy movie be it. 1990s Serbian production with Orson Wells as J.P. Morgan. A pirated version is on You-tube under another title.

Tesla: Inventor of the Industrial Age
by Bernie Carlson
Princeton Univ. Press, 2013
The most recent Tesla biography is a scholarly work, refreshingly free of The Mystique but weak on the technology, like all biographies. Professor Carlson, who must coexist with the physics department down the hall, insists that Tesla's radio and wireless power does not work.

Wizard
by Marc Seifer, 1997
All new research is welcome, including this manifestation of the Mystique.

Prodigal Genius
by John O'Neal, 1946
The Mystique prototype is still read.

My Inventions
21st Century Books
Tesla's autobiography.

Edison
by Matthew Josephson
McGraw-Hill
A biography of Tesla's rival.

other literature

200 Meters and Down:
The Story of Amateur Radio
by Clinton DeSoto, 1936
ARRL, 225 Main St., Newington, CT 06111

U.S. Frequency Allocations
The spectrum, a full-color wall poster chart.
U.S. Govt. Printing Office
Washington, DC 20402

Borderland Sciences
The Journal of Borderland Sciences, back issues, books, Vril Telephony, Gerry Vassalatos, Bolinis, Eric Dollard.
www.borderlands.com

The Lowdown, the LowFER and MedFER
Newsletter, Bill Oliver, ed.
Longwave Club of America
45 Wildflower Rd., Levittown, PA 19057

The Medium Frequency Scrapbook
by Ken Cornell
This bible of LF tech includes some "open land" experiments.

Beacon Guide
by Ken Stryker

The Sounds of Natural Radio
by Michael Mideke

Subsurface Antennas and the Amateur
by Richard Silber

The Antenna Compendium
ARRL

Radios That Work for Free
by K.E. Edwards

Henley's 222 Radio Circuit Designs
1920s vacuum-tube schematics, transmitters and receivers.

Tesla's Work with AC and Wireless
ed. by Tesla researcher Leland Anderson
21st Century books

Secrets of Antigravity Propulsion
by Paul Laviolet
A book, and there is a You-tube lecture. Antigravity as a property of high-voltage caps.

The Multiple Wave Oscillator Handbook
edited by Tom Brown
A 350-page compilation of all sorts of articles on the MWO
and the violet ray, now in its 4th edition.

Borderland Sciences
P.O. Box 6250, Eureka, CA 95502
707-445-2247
www.borderlands.com

The Complex Secret of Dr. T. Henry Moray
by Jorge Resines
Borderland Sciences
Radio and TV circuits from 1928 applied to
Moray's free-energy discoveries.

The Tesla Coil Designer
by Walt Noon
A computer program.
3283 Belvedere, Riverside, CA 92507).

Tesla Coil Secrets
by R.A. Ford

Magnetic Amplifiers, a Lost Technology
of the 1950's
by George Trinkaus
PDF at teslapress.com

Induction Coils
by Lowell and Lorrie, 1904
Borderland Sciences
Includes some coherer technology.

Index

Technical Index

Name Index

www.ingramcontent.com/pod-product-compliance
Lightning Source LLC
Chambersburg PA
CBHW051222200326

41519CB00025B/7209